WALKER'S REMODELING REFERENCE BOOK

Estimating Procedures and Cost Guidelines
for
Residential Remodeling

THIRD EDITION

by
HARRY HARDINBROOK

Frank R. Walker Company
P.O. Box 3180
Lisle, IL 60532
(800) 458-3737
www.frankrwalker.com

We are grateful to readers who call our attention to any errors, typographical or otherwise, that they discover in this book. We also invite constructive criticism and suggestions that will make future editions of this book more complete and useful.

Copyright© 1984, 1987, 1998
by
Frank R. Walker Company

Publishers of
Walker's Building Estimator's Reference Book, Walker's Insulation Techniques and Estimating Handbook, Walker's Manual for Construction Cost Estimating, Walker's Quantity Surveying and Basic Construction Estimating, Practical Accounting and Cost Keeping for Contractors, Business Forms for Contractors, Computer Estimating Software, Scales and Calculators

ALL RIGHTS RESERVED
No part of this publication may be reproduced or transmitted in any form or by any means, electronic or mechanical, including photocopy, recording or any information storage and retrieval system now known or to be invented, without permission in writing from the publisher, except by reviewer who wishes to quote brief passages in connection with a review written for inclusion in a magazine, newspaper or broadcast.

Neither Frank R. Walker Company, the author nor the contributors listed in this book guarantees or warrants the correctness or sufficiency of the information contained in this book, nor do any parties assume any responsibility or liability in connection with the use of this book or of any products, processes or pricing described in this book.

ISBN-0-911592-62-8
Printed in the U.S.A.

TABLE OF CONTENTS

1. How to Use This Book 1
2. Kitchens 5
3. Baths 19
4. Room Conversions 35
5. Room Additions 49
6. Flooring and Stairways 77
7. Drywall, Paneling and Ceiling Tile 107
8. Painting and Wallpaper 139
9. Plumbing 173
10. Electrical 185
11. HVAC 209
12. Roofing, Gutters and Downspouts 223
13. Siding and Exterior Painting 251
14. Windows and Doors 287
15. Porches and Decks 307
16. City Cost Indexes 317
17. Mensuration 321
    Index 335

QUICK REFERENCE CHARTS FOR LABOR FACTORS

1. Rough Framing 74
2. Installing Stairs 104
3. Plastering 132
4. Interior and Exterior Trim 134
5. Interior Millwork 135
6. Insulation 248
7. Roofing 249
8. Sheet Metal 250

Photo Credits

We wish to express our appreciation to the individuals and organizations that provided us with photographs appearing on the following pages:

Andersen Window Company 38, 289
Galaxie Builders, Chicago, IL 51, 254, 255, 293
Stanley Door Systems, Troy, MI 95, 296

Chapter 1

HOW TO USE THIS BOOK

In recent years remodeling has been the fastest growing sector of the residential housing market, and the growth has been explosive. Annual growth used to be in the millions of dollars. Today, we are talking billions.

In a book of estimating cost data that is used throughout the world, it is impossible to quote material prices and labor costs that apply universally. Some computations are required on the part of the reader to make this book an accurate estimating reference. Labor costs in this book are developed by first dividing hours worked by the production rate to find the hours per unit.

Hours ÷ Rate = Labor Hours Per Unit

For example, a carpenter should be able to frame and erect from 375 to 425 b.f. (board feet) of lumber per 8-hour day at the following labor cost per 1,000 b.f. Let's assume that your carpenter can do the larger amount, 425 b.f. To find the labor hours per unit, in this case per board foot, divide 8 by 425 which equals 0.0188 hours.

8 (hours) ÷ 425 (rate) = 0.0188 hours per board foot

To find out how long it takes to do 1,000 b.f., multiply 1,000 by 0.0188. It will take 18.8 hours to install 1,000 b.f.

Standard Estimating Tables

The standard estimating tables show the cost per unit. The following example shows three different board foot unit costs—per thousand board foot, per hundred board foot and per board foot:

	Hours	Rate	Total	Rate	Total
Carpenter (1,000 b.f.)	18.8	--	--	$25.60	$481.28
Cost per 100 b.f.		--	--		48.13
Cost per b.f.					0.48

In all of the cost tables we supply an hourly wage rate, by which the man-hours are multiplied to find the total cost and then cost per unit. This hourly rate is a national average for the given trade and may or may not accurately reflect your wage scales.

For the most accurate estimate, there are blank rate and total columns. Insert your own wage rates and figure your own unit prices. Assume the wage rate for carpenters in your area is $19.81. Your costs would be $0.02 less per board foot, which might add up to a significant difference:

	Hours	Rate	Total	Rate	Total
Carpenter (1,000 b.f.)	18.8	$24.00	$451.20	$25.60	$481.28
Cost per 100 b.f.		--	45.12	--	48.13
Cost per b.f.		--	0.45	--	0.48

Labor productivity for an item is often given as a range, because crews and job conditions vary. Costs on the table are based on the average productivity of this range. For example, if the amount of material placed ranges from 1,700 to 2,200 sq. ft., the figure used to develop pricing is the average, 1,950 sq. ft.

When decimals are rounded, the general rule is that 5 or greater is rounded up, less than 5 is rounded down. For example; $0.0565 is round off to $0.06; $0.0532 is round off to $0.05.

The hourly wage rates shown on the cost tables in this book are national averages based on our surveys of current prevailing wages in localities all over the U.S. Figures are combined wage and fringe benefits. They do not include workers compensation, overhead and profit and the like.

The wage rates that we use are quite likely different from your own, and you must make your own calculations on the tables. When used correctly, the cost tables can save you time and money, but for the most accurate estimate, you should figure your own unit prices using your own wage rates and material prices.

Quick Reference Charts for Labor Costs

The Quick Reference Charts, which are found at the end of key chapters in this book, are a fast and convenient way to figure labor costs for many types of work. The experienced estimator knows that the productivity of workers is influenced by many conditions—job site and weather conditions, quantities of work involved, size and mix of crews and experience of workers. Eventually, the estimator must develop his or her own cost data, based on field experience and the company's cost records. The cost data in this book serves as a starting point for the beginner and a quick reference for the established estimator. But

HOW TO USE THIS BOOK

any publication of this type is based on average job conditions. Production rates in a given situation could be higher or lower.

Once you have estimated material quantities, the Quick Reference Charts will help you to figure labor costs quickly and efficiently. Each chart lists the types of work, the unit of measure used for pricing and a multiplying factor, which represents the labor production rate per unit.

Let's say that you have figured the board feet of lumber required for wall studs in a room addition. Once you have determined the number of units of a material item needed, you want to know the labor cost of setting the studs. Turn to *Quick Reference Chart #1* at the end of *Chapter 5, Room Additions*. The second line of the chart gives the labor factor for studs:

Item Description	Unit	Factor
Studs	BF	0.02000

The unit of measure is board feet (BF). The multiplying factor is 0.02000, which represents the man-hours required to set 1 b.f. of studs. Let's say that you figured 500 b.f. of lumber are needed. To determine the labor cost, multiply:

Total board feet x factor x hourly labor rate = labor cost
500 x 0.02 x $25.60 = $256.00

The hourly labor rate should be an average rate. For example, the work described is to be performed by one carpenter and one helper. The carpenter is paid at the rate of $25.60 per hour. The helper is paid $21.50 per hour. You need to find the average hourly rate. The average rate is figured by totaling the sum of all wage rates and dividing by the number of workers. In the example:

(25.60 + 21.50) ÷ 2 = 47.10 ÷ 2 = $23.55 average hourly rate

A list of Quick Reference Charts appears at the front of the book after the Table of Contents. For a more detailed breakdown of the components of labor cost, refer to the cost tables given in the text, the ones described at the beginning of this chapter.

Chapter 2

KITCHENS

BASIC DESIGN

The kitchen is often the focal point of the home. Certainly, it is the focus of concern for most remodeling contractors. Some might argue the importance of the recreation room or the bathroom, but kitchens are a good starting point, because kitchen remodeling raises a number of general issues.

The major advances in kitchens in recent years have been in design and appliances. Kitchens are opening up and flowing into adjacent rooms. Customers are looking for convenience and work-saving appliances like garbage compactors and built-in barbecues. When it comes to design, selecting and placing functional elements for efficient use as well as aesthetic appeal will make a kitchen successful.

When architects or designers are involved, placement is their responsibility, but even so, the contractor should be aware of basic kitchen design, if for no other reason than to answer questions or make suggestions.

The Work Triangle. Central to the design and function of all kitchens are three work centers: the sink, the refrigerator and the range. Together these three elements form what is called the *work triangle*.

Ideally, the total length of the work triangle is not more than 22 feet or less than 13 feet, and the minimum length of a leg of the work triangle is no less than 4 feet. Keeping within those parameters will save your client a lot of unnecessary steps and make meal preparation a more pleasurable operation.

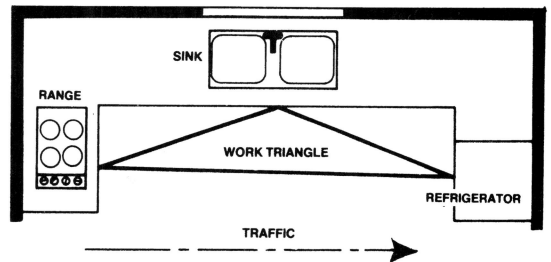

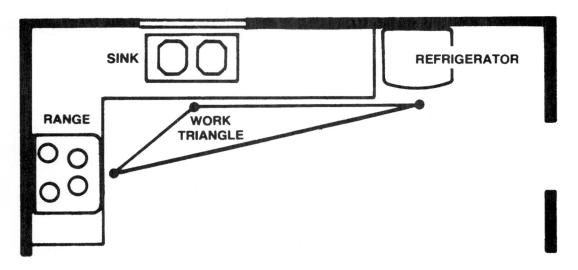

In some homes the kitchen is more than just a place to prepare meals. For some it is also a recreation or family room, and the work triangle should be planned so as to minimize traffic interference from these other activities. Consider whether doors should open from the left or right. Do they interfere with traffic or the work triangle? Adequate counter space should be allocated around each of the work centers, or otherwise, meal preparation and clean up cannot be carried out efficiently.

Care should be taken to allow sufficient space between walls and counters to open drawers without having to step aside and permit another person to pass by. For maximum efficiency the pantry, refrigerator and storage cabinets should be close to the entry door— garage door, side door or back door— where groceries are brought in.

To be sure that food brought to the table is as hot as possible, the range, oven and microwave should be close to the family room table, breakfast nook or dining room. The refrigerator door should swing to the proper side (both right and left swinging models are available) so that food can be conveniently taken out and placed on countertops and returned to the refrigerator. On double-door models it is best to have counter space on both sides of the appliance.

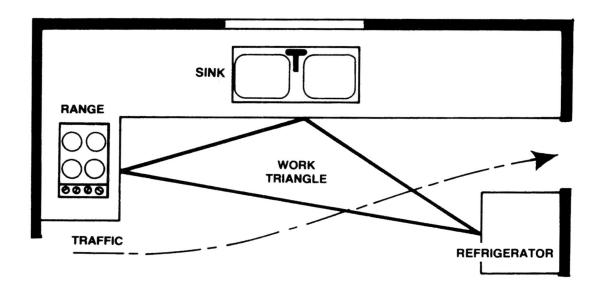

The kitchen work flow is generally from storage or refrigerator to sink for washing or water for cooking to range top or oven, so the sink should be between the refrigerator and stove.

How you resolve these problems will be influenced by the kitchen layout. By giving careful attention to the work triangle and traffic patterns, many headaches are eliminated.

Basic Kitchen Designs. There are four basic kitchen designs, to which you can add peninsulas and islands for any number of variations. By so doing, you can create an attractive and efficient kitchen that will appeal to the most discriminating homeowners. Although each basic design has pluses and minuses, the choice of designs will be influenced to a great extent by the amount and shape of the space available.

Generally, none of these designs is any better than the others, but two of them—the *U-shaped* and the *L-shaped*—provide the greatest flexibility. With these two types, work triangles can be planned so that they have the least traffic interference. Obvious drawbacks are that there might not be sufficient space for a good U-shape, while the L-shape might reduce counter and cabinet space, which affects both the work triangle and the traffic pattern.

Of the other two designs, the *single wall* and the *corridor,* the corridor is most desirable. The big plus is that it can be a real step-saver in the work triangle, and the key cabinets and counters can be near at hand. The biggest drawback for both is the traffic pattern right through the work triangle. A second disadvantage for the single wall layout is lack of space. And obviously, depending on overall space, neither is well suited for entertainment or family activities. The good contractor can work around these limitations to create a kitchen that is both efficient and appealing.

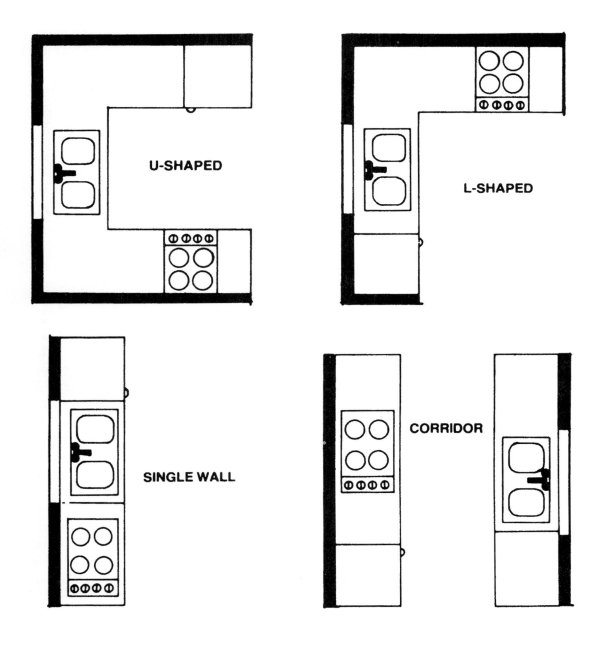

KITCHENS

FUNCTION AND COLOR

After the work triangle has been decided, the next step in efficient kitchen design is to provide for proper storage in a convenient location. Cooking utensils should be near the stove, food preparation utensils near the sink, storage containers near the refrigerator and dining utensils as close as possible to the dining

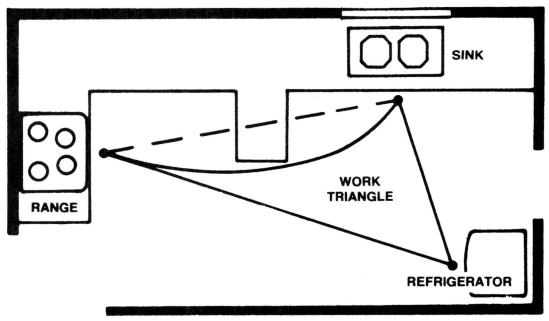

area. This means that placement of cabinets is as important as the type selected.

Where storage space is limited, consideration should be given to using additional cabinets above the wall cases. Custom cabinets can be designed to use the soffit space for added storage area. However, manufacturers of pre-assembled cabinets offer what are known as *short cabinets*. They require a minimum 15" of usable space.

Color is key, and whether it is appliances, wall treatments, cabinets or floor coverings, there is sufficient variety to indulge the taste and personality of every homeowner. Decorators and designers suggest that a single base color be selected, such as brown. It can be the brown in the floor covering, the cabinet facing, counter top or an appliance. Start simple and build up in layers, for example, yellow and beige, to which you can add contrasting brilliant colors as accents. As many as four or five colors can be used which compliment or enhance the base color.

Texture is another element. One manufacturer offers 108 choices in decorative surfacing, ranging from tile to leather, from woodgrain to marble. Counter tops can be in simulated wood, like butcher block, and floor coverings in simulated brick, or cabinet facings in simulated wood like dark hickory and counter tops in simulated leather. Most are low on maintenance and high on aesthetic appeal.

Today's remodeling contractor has a great variety of high quality materials and appliances, which will help justify the costs a contractor charges just to make a profit in today's competitive market.

Even if you have a remodeling contract that calls for redoing only part of a kitchen, you may want to review particular sections in this chapter. For remedial jobs involving rewiring, replumbing or simply cosmetic changes, such projects are covered in more detail in the chapters covering specific trade.

Before commencing work, the contractor should make allowances for the following:

- protection of the adjacent area
- removal of rubbish and salvage
- restrictions during normal working hours
- daily clean-up involved
- availability of plans of original structure showing mechanical and structural details
- checking of building codes to determine if there is work required beyond what has been ordered, such as fire or window exits or strengthening structural members
- visual inspection of appliances and fixtures before fully uncrating or unpacking

REMOVING CABINETS AND APPLIANCES

Sink Removal. The following procedure involves one plumber removing a two-well sink without garbage disposal. Turn off water lines and disconnect drain. Remove trap and drain. Disconnect fixtures and lift sink out.

KITCHENS

	Hours	Rate	Total	Rate	Total
Plumber	1.0	--	--	$29.30	$29.30

If the sink is supported by a wall hanger or joined to the countertop with a trim ring, add 15 minutes to the above time. If a garbage disposal is to be removed, add another 20 minutes.

Installing Shut-off Valves. Each water line requires its own shut-off valve. Where a dishwasher is to be added, the valve will require a dual outlet. Figure one plumber to install them, allowing 1/2 hour per valve.

	Hours	Rate	Total	Rate	Total
Plumber	0.5	--	--	$29.30	$14.65

Removing Range Tops and Ovens. If electric, turn off power at circuit breaker box. If gas, turn off at outside meter or inside shut-off valve. Lift range top out and disconnect gas or electric fittings, being sure power is off. Remove oven faceplate and slide out. Disconnect electric or gas fittings.

	Hours	Rate	Total	Rate	Total
Electrician	0.5	--	--	$29.65	$14.83
Plumber (Gas)	0.5	--	--	29.30	14.65

If this is the only work performed, journeymen will normally charge at least one hour minimum.

Removing Base Cabinets and Tops. Base and wall cabinets are best removed by reversing the order in which they were installed. It is easiest to remove the base cabinets first and then the wall cabinets. To remove standard cabinets or countertops, figure 1/6 hour carpenter time for each lin. ft. of cabinet at the following labor cost per lin. ft.:

	Hours	Rate	Total	Rate	Total
Carpenter	0.17	--	--	$25.60	$4.35

REMOVING TRIM AND FLOORING

Removing Baseboard. One carpenter should be able to remove 200 lin.ft. per hour of one-member baseboard at the following rate per lin.ft.:

	Hours	Rate	Total	Rate	Total
Cost per 100 l.f.	0.5	--	--	$25.60	$12.80
Cost per l.f.			--		$0.13

Removing Sub-flooring. Where existing sub-floors must be removed, a carpenter and laborer can remove 100 sq. ft. in 1.5 hours at the following cost per 100 sq. ft.:

	Hours	Rate	Total	Rate	Total
Carpenter	1.5	--	--	$25.60	$38.40
Laborer	1.5	--	--	21.50	32.25
Cost per 100 s.f.			--		70.65
Cost per s.f.			--		$0.71

Roughing In. Prior to installation of any new fixtures and appliances being relocated, water, drain lines and electrical outlets should be roughed in. For estimating data, see chapters for *Electrical* and *Plumbing*.

No Roughing In. Where no roughing in of water lines and drain is required, a plumber can place and connect a double kitchen sink in 3 hours. For extensive remodeling or where increased electrical service is to be brought into the kitchen, some removal of drywall to expose studs may be necessary. It may be possible to remove only that portion of the wall to be covered by the new base cabinets. However, the amount can vary from job to job. In order to minimize duplication of estimating information, the cost to remove and replace drywall is covered in Chapter 7, *Drywall, Paneling and Ceiling Tile.*

Changing Interior Doors and Openings. Sometimes, creating an additional opening or changing an existing entrance is desirable. For example, smaller kitchens can appear larger by opening a wall and adding a counter on the family room side.

To close off an existing opening, using 2" x 3" or 2" x 4" framing with drywall on both sides, allow a carpenter 4 hours to close the opening. If the opening was where a door previously existed, allow an additional hour to remove the framing.

To cut a new door in a non-load-bearing wall without electrical, heating ducts or plumbing, allow a carpenter 4 hours to cut and frame the opening. However, if any of these services exist, re-routing is difficult to estimate. See Chapter 14, *Windows and Doors,* for door installation and for moving windows.

KITCHENS

HANGING WALL CABINETS

Cabinets generally set the tone for the entire kitchen—colonial, rustic, contemporary. They also set the color pattern. Latches can be friction catch, magnetic catch, touch latch or sliding. They can be made of wood, metal, plastic or even glass in a variety of widths and depths. And whether they are factory or custom built, they come with doors attached ready for installation.

Wall cabinets should be installed before base cabinets for easier installation and to avoid damage due to reaching and dropping. Single-door units measure 9" to 24" wide, double-door units 27" to 48" in three-inch increments. Heights range from 12" to 48". Depths are from 12" to 14". Special depth cabinets are also available for above the refrigerator. Filler strips are used where the distance wall-to-wall is not an even multiple of 3 inches.

Installing Wall Cabinets. Locate studs and draw plumb lines. Check walls for high spots and eliminate where possible. Mark upper and lower position lines and attach temporary support cleats. (Note: if range hood exhausts behind soffit and not through attic, cut wall opening and install ductwork before hanging soffit.) Hang corner cabinets first.

New oven cabinets should be installed as part of the wall cabinets even though they extend to the floor.

Generally, two workers are needed to install wall cabinets, one to align and hold and one to fasten. A carpenter and laborer should be able to place and hang an average 24 lin.ft. of wall cabinets per 8-hr. day at the following cost per lin.ft.:

	Hours	Rate	Total	Rate	Total
Carpenter	0.33	--	--	$25.60	$ 8.45
Laborer	0.33	--	--	21.50	7.10
Cost per l.f.	0.66		--		$15.55

Installing Ductwork Behind Soffit. Use a segment of duct as a template to scribe a wall cutout. In outer wall, bore hole through center from inside. (If outside wall is masonry, use masonry drill.) Cut exterior hole and insert duct cap. Connect interior ducting to duct cap.

For wood, to complete above .... 6 hrs.
For masonry .... 8 hrs.

Labor to Set Cabinet Hardware

Unit	Hrs./Each
Surface bolts	0.05
Offset bolts	0.10
Catches	0.10
Drawer Pulls	0.06
Knobs	0.05
Rim Catches	0.10
Mortise Lock	0.40

SETTING BASE CABINETS AND COUNTERTOPS

The standard height for base cabinets is 36", but this might be lowered to 30", when counters or peninsulas are used for meals, or when a member of the family is in a wheelchair. In fact, it is becoming quite common to lower all base cabinets, including the sink and the range cabinets, for the physically challenged to facilitate meal preparation and minimize accidents that can be potentially serious, particularly around the range. For material costs, see Chapter 4 for lumber and gypsum board.

Installing Base Cabinets. Locate studs and draw plumblines. Check walls for flatness and mark where shims are required to eliminate high spots. Check the floor level and mark where shimming is required. Chalkline wall and floor for position.

A carpenter should be able to place and attach 24 lin. ft. of base cabinets per 8-hour day at the following cost per lin. ft.:

	Hours	Rate	Total	Rate	Total
Cost per l.f.	0.33	--	--	$25.60	$8.45

Installing Countertops. Countertops come in a variety of surfaces, each having advantages and disadvantages. Formica® brand, plastic laminate, is still the most popular, because it does not fade or stain, is easy to maintain and is one of the most economical. Sheet vinyl is the cheapest and easy to clean, but it does not hold up well and marks quite easily. The two with the longest life are stainless steel and tile. Although maintenance is easy, the former is noisy and cold to the look and touch, and with the latter, dishes break easily and joints loosen. Wood is difficult to clean, and it can splinter. When it is covered with an acrylic coating, these problems are eliminated, but the acrylic itself is susceptible to

KITCHENS

marks and cuts. Another countertop is Corian® by DuPont. This product is harder than marble and will not stain and smudge.

Let's take as an example a U-shaped Formica countertop. Using standard section shapes, there will be two left corners (L) and two right corners (R) and no straight (S) sections. Mark the two (L) sections 1 and 3 and the two right (R) sections 2 and 4. When trimming sections to size, any fine-tooth or carbide saw can be used. To avoid ragged edges for cutting with either a circular or saber saw, place decorative face down; when using a table saw or hand saw, place decorative face up. Laminate 1/32" thick can be cut with tin snips.

Cut 1/8" to 1/4" over size and trim after mounting. A special router bit, plane or file is used to trim and bevel corners. Start with sections 1 and 3. Apply contact bond cement on both the laminate and counter with a spreader, roller or brush following manufacturers instructions and specifications. Align front edges and top surface. Tighten joint fasteners front to back. Place sections on cabinet and fasten with wood screws from underneath. Repeat procedure with sections 2 and 4.

A carpenter (long tops, especially L's, require two workers on site) should be able to place and attach 5 lin. ft. of countertop per hour at the following cost:

	Hours	Rate	Total	Rate	Total
Cost per l.f.	0.20	--	--	$25.60	$5.12

Setting Factory Assembled and Finished Kitchen Cabinets. When setting factory assembled and finished kitchen cabinets, such as used in the average apartment or residence, it will require about 0.1 hours of carpenter time for each sq. ft. of cabinet face area at the following labor cost per sq. ft.:

	Hours	Rate	Total	Rate	Total
Cost per s.f.	0.10	--	--	$25.60	$2.56

For *first grade workmanship* in quality apartments and residences, where a high-class job of installation is required, it will take about 1/6 hour carpenter time for each sq. ft. of cabinet face area at the following labor cost per sq. ft.:

	Hours	Rate	Total	Rate	Total
Cost per s.f.	0.16	--	--	$25.60	$4.10

Cutouts. In making sink and range cutouts, use the sink trim ring and the range tub as templates. After marking, cover saw line with clear tape and drill

1/2" holes at opposite corners. Make cutout with fine-tooth saw, using fine-tooth file from top to bottom to smooth edges. Figure a single carpenter one hour per cutout.

However, countertops can be ordered with exact specifications including cutouts from cabinet suppliers. Installation is to simply position and fasten.

Installing Sink, Range (Gas or Electric) and Oven. Figure one worker to do each installation at the following approximate flat rate:

Place and connect sink	$250.00
Place and connect range, electric	66.00
Place and connect range, gas	83.00
Place and connect oven, electric	66.00
Place and connect oven, gas	83.00

SOFFITS, RANGE HOODS, DISPOSALS

Making and Installing Soffits. Soffits are generally made from 2" x 2" or 2" x 4" lumber. This estimate includes cutting, assembling, placing and finishing off with drywall or paneling. A carpenter should be able to install 8 lin. ft. per hour at the following cost per lin. ft.:

	Hours	Rate	Total	Rate	Total
Cost per l.f.	0.125	--	--	$25.60	$3.20

Installing Range Hoods. Let's assume a range hood that is exhausted through the wall cabinet and soffit. This requires two workers. The procedure for installing ductwork behind the soffit was described earlier in this chapter. Position the hood collar below the bottom shelf of the wall cabinet and mark the shelf to show the shape of the duct cutout. Make the appropriate cutouts in the shelves, soffit and the ceiling, if the duct is to go through the roof. For a roof vent, allow for placing a vent cap, flashing where the vent cap enters the roof and replacing shingles for a one-story vent, a kitchen on the second level or a ranch-style. Connect electrical conduit to the junction box, and check the operation of fan and light. Install and attach duct to collar and fasten hood to lower shelf. An electrician and carpenter will need about 7 man-hours total at the following cost:

	Hours	Rate	Total	Rate	Total
Electrician	2.5	--	--	29.65	$ 74.13
Carpenter	4.5	--	--	$25.60	115.20
	7.0				$189.33

A popular alternative is the ductless range hood. This unit filters the air through a grease filter and an activated charcoal filter, then back into the room. Figure the installation will take 4.5 hours at the following cost:

	Hours	Rate	Total	Rate	Total
Electrician	2.5	--	--	$29.65	$ 74.13
Carpenter	2.0	--	--	25.60	51.20
	4.5				$125.33

Installing Garbage Disposal. For a new installation, an outlet and wall switch should have been roughed-in prior to placing the cabinets and sink. For these items, see the chapter on *Electrical.* For safety the wall switch should be positioned 6 feet from the sink centerline. Check disposal for vertical clearance. Apply plumbers putty around the sink drain opening. Insert the sink sleeve from above and the mounting assembly from underneath. Secure with a snap ring and tighten the mounting screws evenly. Attach the discharge tube and insert the disposal into the mounting assembly. Connect the power and check operation. A plumber should be able to complete the installation in 1.5 hours at the following labor cost:

	Hours	Rate	Total	Rate	Total
Plumber	1.5	--	--	29.30	$43.95

Obviously, remodeling an entire kitchen entails considerably more work than has been covered in this chapter. We have tried to cover here the items that are specific to the kitchen. Work such as flooring that pertains not only to the kitchen but to every room in the home is covered in detail in specific chapters for the application or trade.

Typical Kitchen Material Costs

As one can see from the ranges given below, prices for items such as countertops, faucets, sinks and cabinets vary widely, according to style, material and quality. An accurate remodeling estimate depends on deciding with the client exactly what is desired and on obtaining the most current pricing from suppliers. The prices shown can give one a ballpark figure, but they must not be used in place of actual material cost quotes from your suppliers.

Countertops (4', 6', 8', 10', 12' lengths)   **Price**
Laminate countertops .... $4.15–9.15 l.f.
DuPont Corian® .... 97.00 l.f.

Kitchen Faucets
Washerless (without spray) 39.00
Washerless (with spray) 59.00–150.00

Kitchen Sinks
Porcelain on steel (double-well, 33" x 22")
    White $168.00
    Color 178.00
Stainless Steel (double-well, 33" x 22") 40.00–85.00
Faucets 39.00–79.00
Garbage disposal 50.00–100.00

Wall Cabinets
Based on standard L-shaped 20 lineal-foot kitchen. Estimated total for wall and base cabinets.

White melamine surface .... $ 800.00
Matte white thermo-foil surface 1050.00
Oak, veneer center panel 1200.00
Oak, solid center panel 1400.00
Maple, veneer center panel 1200.00
Solid hickory hardwood 2000.00

Chapter 3

BATHS

Bathrooms come in a wide range of styles, from the very simple to the luxurious. Among the high-end options are whirlpool baths and hot tubs, his and her lavatories, shampoo lavatories, bidets and starkly modern or traditionally-designed baths, toilets and lavatories.

When it comes to adding bathroom facilities, extra-small baths and half-baths can be installed in spaces as small as 4' x 4', which is room enough for a W.C. and lavatory.

Fixtures such as shower stalls are available in 32" x 32" or 36" x 36" in enameled steel and seamless plastic. Even tubs come in small sizes. For example, 14" high x 42" long x 34" wide and 39" long x 38" wide are just a couple of the available sizes. There are corner tubs, sunken tubs, raised tubs and tubs big enough for two, three or even four people.

Toilets also come in many different styles. There are low styles for invalids and new one-piece designs with integrated water tanks suitable for almost any size bath. Bidets, long a European tradition, are found in American homes. Not only do they come in a variety of colors and styles, they also come in sets so that the bidet and the toilet are a compatible style and color.

As for lavatories the variety of sizes and styles is practically unlimited. A contractor can find one that fits the decor and space of almost any type of bathroom design.

Even such utilitarian items as bath cabinets are being designed for style, and they are being designed with the remodeling contractor in mind. The trend in recent years is surface-mounted rather than recessed cabinets. For years there has been no standard opening size for recessed ones. There have been two general geographically-defined standards. In the East through the Midwest to Denver, the most common size is 14"x18"; from Denver west to the Coast, it is 14"x24". However, there are many variations in between, and errors in size openings and location occur with great frequency.

Changing construction techniques have influenced installation and style of bath cabinets as well. With bathrooms being installed back-to-back, there just isn't room to recess the cabinets back-to-back. There is always the contractor's nemesis, "What's behind that wall?" Is it really worth cutting a hole in the wall for a $15.00 item only to discover that because of wiring, plumbing or ductwork, it will take another $500.00 in labor and materials to complete the installation?

Homeowners want the maximum amount of mirror and the maximum amount of storage. Surface-mounted units provide both.

Swinging doors are popular, and wood, particularly oak, is the leader. It is also interesting to note that for every free-standing tub or lavatory a contractor removes to replace with built-ins, just as many built-ins are removed to be replaced with free-standing fixtures. High style is making its mark everywhere, and bathrooms are no exception.

When doing bathroom remodeling, it is best to complete the bathtub installation first. Many experts suggest not putting the tub or shower against an outer wall, unless the wall and tub are properly insulated and have a vapor barrier. Make sure the fan is vented to the outside and not to the attic space. Otherwise, moisture will reach it's dew point and condense in the attic area, or water can condense on the outside of the tub or shower wall and run down the wall cavity or seep into the flooring eventually causing deterioration.

TOILETS

Removing Existing Toilet. *Floor-mounted unit.* A floor-mounted unit can usually be removed by a single plumber. The procedure takes into consideration the following: draining toilet, disconnecting water supply lines, cutting or otherwise removing stubborn bowl mounting bolts, cleaning old setting compound from closet flange and floor and removing components from home for disposal. A plumber can do it in two hours. *Wall-mounted unit.* The procedure is basically the same. Figure two hours labor time.

Installing Shut-off Valves. Many older homes lack shut-off valves, and a new valve should be added before the new toilet is installed. Using flexible connector, figure 0.5 hours for the installation. For wash basins, simply double the preceding.

Installing New Toilet. *Floor-mounted unit.* When installing a floor-mounted unit to existing waste pipe and water supply, figure one plumber can complete the installation in 2.5 hours. *Wall-mounted unit.* This will require a helper. To the above, add 1 hour for helper time.

Installing a Bidet. The installation procedure is similar to that of floor-mounted toilets, except that it requires a hot water supply as well as cold and uses a smaller drain. However, the big cost in adding a bidet comes in installing a new waste pipe, cutting into the stack and adding new water supply lines. (See below for roughing in.) Bidets do not require water closets. Figure one plumber can install and hook up the bowl in 4 hours.

BATHS

Roughing In. Where new waste pipe and water lines are required or a bathroom appliance is to be moved to a different location and roughing in is required, see Chapter 9 *Plumbing*.

BATHTUBS

Removing Existing Tubs. *Free-standing.* Installations of this kind seldom pose any unusual problems, but because of their weight, bathtubs usually require two workers for removal. Allow 2 man-hours for a plumber and 2 man-hours for a helper to disconnect the existing fittings and pipes and remove the tub. *Built-in or recess tubs.* Add 2 man-hours to the preceding.

Roughing in New Plumbing Connections. Older free-standing units are often replaced with built-ins that require new plumbing. Generally, it requires a new elbow and strainer stopper mechanism, overflow elbow and plate, waste pipe and hot and cold water lines. Allow one plumber 4 hours for this procedure.

Installing New Tub/Shower. If new framing is required, add 2 hours for one carpenter to rough in the necessary studs and tub supports on three walls (if building code requires). Figure one plumber can do the necessary hookups, including new fittings and installing the shower head, in 1 hour. If removing or adding a tub or shower unit, make sure the opening is large enough for the unit to pass through! Add costs for removal of doors, opening of wall and door replacement if necessary.

CABINETS/VANITIES/WASH BASINS/ACCESSORIES

Wall Cabinets. If an old wall cabinet exists, allow 1 hour for its removal. If no cabinet exists, it will be necessary to cut and frame a recess for the new cabinet. Allow a minimum of 1.5 hours for this procedure.

Lighted Cabinets. Many of today's cabinets have self-contained lighting. Figure one electrician can complete the necessary hook up from an existing 110-volt source in 1.5 hours.

Fitting and Placing Closet Shelving. A carpenter should fit and place 115 to 135 sq. ft. of closet shelving per 8-hour day (including setting of shelf cleats) at the following labor cost per 100 sq. ft.:

	Hours	Rate	Total	Rate	Total
Carpenter	6.4	--	--	$25.60	$163.84
Cost per s.f.			--		$1.64

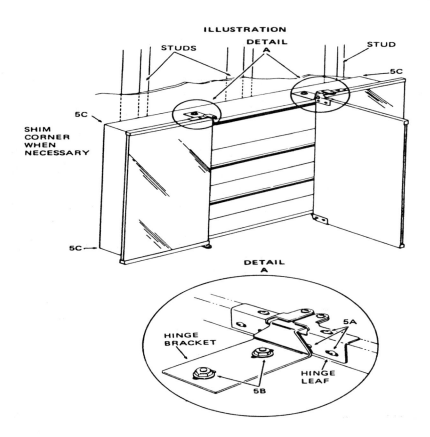

Installing Hanging Rods in Closets. A carpenter should install about three hanging rods, including supports, per hour at the following labor cost per rod:

	Hours	Rate	Total	Rate	Total
Carpenter	0.33	--	--	$25.60	$8.45

Setting Bathroom Accessories. In an 8-hour day a carpenter should locate and set 30 to 34 bathroom accessories such as towel bars, soap dishes and paper holders at the following labor cost per accessory:

	Hours	Rate	Total	Rate	Total
Carpenter	0.25	--	--	$25.60	$6.40

BATHS

Removing Lavatory or Wash Basin. To remove a wall-hung lavatory, figure one worker can disconnect the waste and water lines and remove the fittings in 1 hour. (If new shut-off valves are required, add 0.5 hours per valve.)

Removing Vanities. The removal of kitchen cabinets, vanities and built-in cabinets is figured by the lineal feet of facing. Figure 0.167 hours carpenter time for each lin.ft. of vanity. Add for removal of debris and clean up.

Installing Vanity. Most vanities today come pre-assembled ready to install. Figure a carpenter can install a standard 24" built-in vanity in 2 hours.

Installing Wall-Hung Lavatory. Modern bathrooms come in many configurations, including wall-hung lavatories. Installing lavatories integrated with countertop, lavatories from above and lavatories from below all require 2 hours.

Ceiling Ventilators. Installing through-the-ceiling ventilators involves cutting an opening through the ceiling, close to a joist and large enough for the ventilator housing. The housing is attached to the joist. The fan motor assembly is screwed into the housing and a grid shield is attached to conceal the unit and the opening. Operation is by wall switch. The labor cost for an electrician to install:

	Hours	Rate	Total	Rate	Total
Open Ceiling	3.75	--	--	$29.65	$111.19
Closed Ceiling	5.25	--	--	29.65	155.66

Wall Ventilators. A typical through-the-wall ventilator consists of an outer sleeve inserted into the wall from outside the house, an inner sleeve forced into the outer sleeve from inside and a fan-motor assembly. Openings are cut through both inside and outside walls. The fan-motor assembly is attached to the inner sleeve. A concealment grill is placed on the inside and a hinged door on the outside to prevent backdrafts. Operation is also by wall switch. Labor cost for an electrician to install:

	Hours	Rate	Total	Rate	Total
Open Wall	4.75	--	--	$29.65	$140.84
Closed Wall	6.25	--	--	$29.65	185.31

If the exterior finish is stucco, add 1 hour; if brick, add 2 hours.

AUXILIARY HEATERS

Ceiling Heaters. Most electrical ceiling heaters are flush-mounted to a mounting plate and secured to a junction box fastened to a ceiling joist. The unit is operated by wall switch. Labor cost for an electrician to install:

	Hours	Rate	Total	Rate	Total
Open Ceiling	3.75	--	--	$29.65	$111.19
Closed Ceiling	5.25	--	--	29.65	155.66

Wall Heaters. Wall heaters are recessed into the wall and fastened to the nearest stud. Operation is by wall switch and electrical power is brought to the unit via conduit from an existing circuit. Labor cost for an electrician to install:

	Hours	Rate	Total	Rate	Total
Open Wall	2.5	--	--	$29.65	$ 74.88
Closed Wall	4.0	--	--	$29.65	118.60

Baseboard Heaters. Baseboard heaters are surface-mounted at the base of a wall. Like ceiling and wall heaters, a power source must be brought to the heater and activated by a wall switch. Labor cost for an electrician to install:

	Hours	Rate	Total	Rate	Total
Open Wall	1.5	--	--	$29.65	$44.48
Closed Wall	2.5	--	--	$29.65	74.13

WALL FINISHES

Alkyd Paint. This is the most economical type of finish. Alkyd (solvent-thinned) paint is flammable, but it will bond when applied over latex, on bare wood and on wallpaper with proper surface preparation. It does not bond well to bare plaster, unprimed wallboard or bare masonry.

Latex Paint. This water-thinned, non-flammable paint does not bond well to alkyd paint, wallpaper or new wood. Latex can be used over new plaster, cinder block and unprimed wallboard.

Urethanes and Epoxies. These require special solvents for thinning and clean-up. Highly resistant to abrasion, moisture and household chemicals, they are very durable. Urethanes adhere well to old paints, including latex and alkyd, and can be used on bare wood as primer sealer and finish. Epoxies are ideal for non-porous materials like tile and glass and even concrete. However, they do not adhere well to previously painted surfaces. For estimating data on interior painting, see Chapter 8, *Painting and Wallpaper.*

WALLPAPER

Vinyls. The fabric-backed type are best and easiest to work with. The next best is vinyl-coated paper or paper-backed vinyl. For estimating data, see Chapter 8, *Painting and Wallpaper.*

Metallic Foils. These are good, provided they are backed with fabric or paper. They are somewhat difficult to work with compared with vinyl.

Prefinished Panels. These come in a variety of finishes, patterns and textures. Some resemble ceramic tile in appearance and touch. They are applied directly to smooth existing surfaces with adhesives.

Others come in rich wood grains protected with coatings to resist heat and humidity and are nailed to studs or framing strips.

TILE

Ceramic is still the best finish for bathroom walls. It has a long life and is easy to maintain, and tile wainscoting is still a practical and decorative favorite.

Bathroom floors and ceilings can be finished in any of a wide variety of materials, but the material must be moisture-resistant. In fact, many of the materials used in other areas of the house are not only suitable for the bathroom but are even desirable because of the decorative options they offer. For floors there are woods, strip flooring, parquet and plank, quarry and mosaic tile, carpeting, sheet vinyl and vinyl tile. For ceilings there are tile and suspended ceilings.

Ceramic Tile Bathroom Accessories. These accessories are available in a variety of sizes, colors and styles, recessed or surface-mounted:

Description	per piece
Recessed soap holder	$12.75
Recessed glass holder	12.75
Roll paper holder	15.90
Robe hook	4.25
Double robe hook	4.75
Toothbrush holder	6.36
Tile bar brackets (pair)	8.50

Decorative Ceramic Wall Tiles. The designs of glazed ceramic wall tile are an integral part of each tile. This tile is normally supplied in 4-1/4" x 4-1/4" and 6" x 6" sizes. The price of this material varies with the type of and number of colors in the design. If the design is known, the estimator should obtain a firm price from a supplier. A general price range is:

		per s.f.
One Color	4-1/4" x 4-1/4"	$1.04–2.65
	6" x 6"	1.20–2.50

Ceramic Tile Adhesives and Accessory Materials

Item	Coverage	Price
Organic adhesive	40-50 sf/gal	$8.50/gal
Dry-set mortar mix	35 lbs/100 sf	0.50/lb.
Primer (for use in damp areas)	125 sf/gal	6.50/gal.
Plastic underlayment	100 sf/unit*	8.50/unit
Wet tile grout	15-20 sf/lb**	0.30/lb.
Dry tile grout	12-15 sf/lb**	0.35/lb.

*Unit consists of 1 gallon liquid and 34 lbs. aggregate.
**When used with 4-1/4" x 4-1/4" tile.

Cost of 100 sq. ft. of 4-1/4" x 4-1/4" Glazed Ceramic Wall Tile Conventional Mortar Method of Installation (unmounted tile)

	Hours	Rate	Total	Rate	Total
2 sacks portland cement	--	--	--	$5.66	$ 11.32
4 cu. ft. sand	--	--	--	$0.53	2.12
100 sq. ft. glazed wall tile	--	--	--	1.99	199.00
10 lbs. dry tile grout	--	--	--	--	9.99
Tile setter	13.5	--	--	25.60	345.60
Cost per 100 s.f.			--		$568.03
Cost per s.f.			--		5.68

BATHS

Cost of 100 sq. ft. of 4-1/4" x 4-1/4" Glazed Ceramic Wall Tile
Water Resistant Organic Adhesive (unmounted tile)

	Hours	Rate	Total	Rate	Total
2.5 gallons adhesive	--	--	--	$15.99	$ 39.98
100 sq. ft. glazed wall tile	--	--	--	1.99	199.00
10 lbs. dry tile grout mix	--	--	--	--	9.90
Tile setter	8.0	--	--	25.60	204.80
Cost per 100 s.f.			--		$453.68
Cost per s.f.			--		4.54

Cost of 100 sq. ft. of 4-1/4" x 4-1/4" Glazed Ceramic Wall Tile
"Dry-Set" Portland Cement Mortar (unmounted tile)

	Hours	Rate	Total	Rate	Total
35 lbs. dry-set mortar mix	--	--	--	$0.32	$ 11.20
100 sq. ft. glazed wall tile	--	--	--	1.99	199.00
10 lbs. dry tile grout mix	--	--	--	--	9.99
Tile setter	10.0	--	--	25.60	256.00
Cost per 100 s.f.			--		$476.19
Cost per s.f.			--		4.76

Cost of 100 sq. ft. of 4-1/4" x 4-1/4" Glazed Ceramic Wall Tile
Conventional Mortar Method of Installation (back-mounted tile)

	Hours	Rate	Total	Rate	Total
2 sacks of portland cement	--	--	--	$5.66	$ 11.32
4 cu. ft. sand	--	--	--	0.53	2.12
100 sq. ft. glazed wall tile	--	--	--	1.99	199.00
10 lbs. dry tile grout mix	--	--	--	--	9.99
Tile setter	12.0	--	--	18.75	307.20
Cost per 100 s.f.			--		$529.63
Cost per s.f.			--		5.30

Cost of 100 sq. ft. of 4-1/4" x 4-1/4" Glazed Ceramic Wall Tile
Water Resistant Organic Adhesive (back-mounted tile)

	Hours	Rate	Total	Rate	Total
2.5 gallons adhesive		--	--	$15.99	$39.98
100 sq. ft. glazed wall tile		--	--	1.99	199.00
10 lbs. dry tile grout mix		--	--	--	9.99
Tile setter	8.0	--	--	25.60	204.80
Cost per 100 s.f.				--	$453.77
Cost per s.f.				--	4.54

Cost of 100 sq. ft. of 4-1/4" x 4-1/4" Glazed Ceramic Wall Tile
"Dry-Set" Portland Cement Mortar (back-mounted tile)

	Hours	Rate	Total	Rate	Total
35 lbs. dry-set mortar mix		--	--	$0.32	$ 11.20
100 sq. ft. glazed wall tile		--	--	1.99	199.00
10 lbs. dry tile grout mix		--	--	--	9.99
Tile setter	9.0	--	--	25.60	230.40
Cost per 100 s.f.				--	$450.59
Cost per s.f.				--	4.51

Note: Add for plastering when specified.

Cost of 100 sq. ft. of 1" x 1" Ceramic Mosaic Tile Floors
Conventional Mortar Method of Installation (face-mounted tile)

	Hours	Rate	Total	Rate	Total
2 sacks portland cement		--	--	$5.66	$ 11.32
4 cu. ft. sand		--	--	0.53	2.12
100 sq. ft. ceramic mosaic tile		--	--	1.69	169.00
25 lbs. dry tile grout mix		--	--	--	24.98
Tile setter	10.5	--	--	25.60	268.80
Cost per 100 s.f.				--	$476.22
Cost per s.f.				--	4.76

BATHS

Cost of 100 sq. ft. of 1" x 1" Ceramic Mosaic Tile Floors
Water Resistant Organic Adhesive (face-mounted tile)

	Hours	Rate	Total	Rate	Total
2.5 gallons adhesive		--	--	$15.99	$ 39.98
100 sq. ft. ceramic mosaic tile		--	--	1.69	169.00
25 lbs. dry tile grout mix		--	--	--	24.98
Tile setter	8.0	--	--	25.60	204.80
Cost per 100 s.f.				--	$438.76
Cost per s.f.				--	4.39

Cost of 100 sq. ft. of 1" x 1" Ceramic Mosaic Tile Floors
"Dry-Set" Portland Cement Mortar (face-mounted tile)

	Hours	Rate	Total	Rate	Total
35 lbs dry-set mortar mix		--	--	$0.32	$ 11.20
1 cu. ft. of sand		--	--		0.53
100 sq. ft. ceramic mosaic tile		--	--	1.69	169.00
25 lbs. dry tile grout mix		--	--		24.98
Tile setter	9.0	--	--	25.60	230.40
Cost per 100 s.f.				--	$436.11
Cost per s.f.				--	4.36

Note: Add for cement floor fill or plastic underlayment when required.

Cost of 100 sq. ft. of 1" x 1" Ceramic Mosaic Tile Floors
Conventional Mortar Method of Installation (back-mounted tile)

	Hours	Rate	Total	Rate	Total
2 sacks portland cement		--	--	$5.69	$ 11.32
4 cu. ft. of sand		--	--	0.53	2.12
100 sq. ft. ceramic mosaic tile		--	--	1.69	169.00
25 lbs. dry tile grout mix		--	--		24.98
Tile setter	11.0	--	--	25.60	281.60
Cost per 100 s.f.				--	$489.02
Cost per s.f.				--	4.89

Cost of 100 sq. ft. of 1" x 1" Ceramic Mosaic Tile Floors
Water Resistant Organic Adhesive (back-mounted tile)

	Hours	Rate	Total	Rate	Total
2.5 gallons adhesive................................		--	--	$15.99	$ 39.98
100 sq. ft. ceramic mosaic tile..............		--	--	1.69	169.00
25 lbs. dry tile grout mix......................		--	--		24.98
Tile setter...	8.0	--	--	25.60	204.80
Cost per 100 s.f.			--		$438.76
Cost per s.f.			--		4.39

Cost of 100 sq. ft. of 1" x 1" Ceramic Mosaic Tile Floors
"Dry-Set" Portland Cement Mortar (back-mounted tile)

	Hours	Rate	Total	Rate	Total
35 lbs dry-set mortar mix.....................		--	--	$0.32	$ 11.20
1 cu. ft. sand...		--	--		0.53
100 sq. ft. ceramic mosaic tile..............		--	--	1.69	169.00
25 lbs. dry tile grout mix......................		--	--		24.98
Tile setter...	9.5	--	--	25.60	243.20
Cost per 100 s.f.			--		$448.91
Cost per s.f.			--		4.49

Note: Add for cement floor fill or plastic underlayment when required.

Cost of 100 sq. ft. of 1" x 1" Ceramic Mosaic Tile on Walls
Conventional Mortar Method of Installation (face-mounted tile)

	Hours	Rate	Total	Rate	Total
2 sacks of portland cement...................		--	--	$5.66	$ 11.32
4 cu. ft. sand...		--	--	0.53	2.12
100 sq. ft. ceramic mosaic tile..............		--	--	1.69	169.00
25 lbs. dry tile grout mix......................		--	--		24.98
Tile setter...	11.0	--	--	25.60	281.60
Cost per 100 s.f.			--		$489.02
Cost per s.f.			--		4.89

BATHS

Cost of 100 sq. ft. of 1" x 1" Ceramic Mosaic Tile on Walls
Water Resistant Organic Adhesive (face-mounted tile)

	Hours	Rate	Total	Rate	Total
2.5 gallons adhesive	--	--	--	$15.99	$ 39.98
100 sq. ft. ceramic mosaic tile		--	--	1.69	169.00
25 lbs. dry tile grout mix		--	--		24.98
Tile setter	9.0	--	--	18.75	230.40
Cost per 100 s.f.			--		$464.36
Cost per s.f.			--		4.64

Cost of 100 sq. ft. of 1" x 1" Ceramic Mosaic Tile on Walls
"Dry-Set" Portland Cement Mortar (face-mounted tile)

	Hours	Rate	Total	Rate	Total
35 lbs. dry-set mortar mix		--	--	$0.32	$ 11.20
1 cu. ft. sand				--	0.53
100 sq. ft. ceramic mosaic tile		--	--	1.69	169.00
25 lbs. dry tile grout mix		--	--		24.98
Tile setter	9.0	--	--	25.60	230.40
Cost per 100 s.f.			--		$436.11
Cost per s.f.			--		4.36

Note: Add for plastering when specified.

Cost of 100 sq. ft. of 1" x 1" Ceramic Mosaic Tile on Walls
Conventional Mortar Method of Installation (back-mounted tile)

	Hours	Rate	Total	Rate	Total
2 sacks portland cement		--	--	$5.66	$ 11.32
4 cu. ft. sand		--	--	0.53	2.12
100 sq. ft. ceramic mosaic tile		--	--	1.69	169.00
25 lbs. dry tile grout mix		--	--		24.98
Tile setter	10.5	--	--	25.60	268.80
Cost per 100 s.f.			--		$476.22
Cost per s.f.			--		4.76

Cost of 100 sq. ft. of 1" x 1" Ceramic Mosaic Tile on Walls
Water Resistant Organic Adhesive (back-mounted tile)

	Hours	Rate	Total	Rate	Total
2.5 gallons adhesive		--	--	$15.99	$ 39.98
100 sq. ft. ceramic mosaic tile		--	--	1.69	169.00
25 lbs. dry tile grout mix		--	--		24.98
Tile setter	9.0	--	--	25.60	230.40
Cost per 100 s.f.			--		$464.36
Cost per s.f.			--		4.64

Cost of 100 sq. ft. of 1" x 1" Ceramic Mosaic Tile on Walls
"Dry-Set" Portland Cement Mortar (back-mounted tile)

	Hours	Rate	Total	Rate	Total
35 lbs. dry-set mortar mix		--	--	$0.32	$ 11.20
1 cu. ft. sand					0.53
100 sq. ft. ceramic mosaic tile		--	--	1.69	169.00
25 lbs. dry tile grout mix		--	--		24.98
Tile setter	9.0	--	--	25.60	230.40
Cost per 100 s.f.			--		$436.11
Cost per s.f.			--		4.36

Note: Add for plastering when specified.

Cost of 100 sq. ft. of 6" x 6" x 1/2" Quarry Tile Floors
Conventional Mortar Method of Installation

	Hours	Rate	Total	Rate	Total
2 sacks portland cement		--	--	$5.66	$ 11.32
6 cu. ft. sand		--	--	0.53	3.18
100 sq. ft. quarry tile		--	--	1.56	156.00
35 lbs. portland cement grout		--	--		3.50
Tile setter	16.0	--	--	25.60	409.60
Cost per 100 s.f.			--		$583.60
Cost per s.f.			--		5.84

BATHS

Cost of 100 sq. ft. of 6" x 6" x 1/2" Quarry Tile Floors "Dry-Set" Portland Cement Mortar

	Hours	Rate	Total	Rate	Total
35 lbs. dry-set mortar mix	--	--	--	$0.32	$ 11.20
2 cu. ft. sand	--	--	--	0.53	1.06
100 sq. ft. quarry tile	--	--	--	1.56	156.00
35 lbs. portland cement grout	--	--	--		3.50
Tile setter	12.0	--	--	25.60	307.20
Cost per 100 s.f.			--		$478.96
Cost per s.f.			--		4.79

Placing Cement Floor Fill. Cement floor fill under ceramic tile floors is usually placed by one tile setter and one or two helpers working together. The fill is placed one or two days in advance of the tile if the overall thickness from rough floor to finished tile surface is over 3" thickness and under, fill and setting bed may be placed in one operation. When such tile floors are to be placed over wood subfloors, it is necessary to first place a layer of waterproof building paper and a layer of wire mesh reinforcing before placing the fill.

A tile setter and two helpers should place 450 to 500 sq. ft. of fill per 8-hour day in areas large enough to permit efficient operations at the following cost per 100 sq. ft.:

	Hours	Rate	Total	Rate	Total
6 bags portland cement	--	--	--	$5.66	$ 33.96
1 cu. ft. of sand	--	--	--		14.25
Tile Setter	1.5	--	--	25.60	38.40
Helper	3.0			21.50	64.50
Cost per 100 s.f.	4.5		--		$151.11
Cost per s.f.					1.51

Labor Factors for Small Quantities

		Labor Multiplier
Ceramic Mosaic Tile	Small rooms	1.75
	Countertops	3.00
Glazed Ceramic Wall Tile	Small rooms	1.50
	Mantel fronts	2.00

		Labor Multiplier
	Mantel front w/returns	1.90
Cove or Base	Small rooms	1.25
Cap	Small rooms	1.10

For small jobs multiply the labor costs given in the applicable unit price development by the appropriate labor factor. Material costs will also change.

Typical Bathroom Material Costs

Bath Hardware
    Rope hook, chrome .... $ 5.00–15.00
    Tissue roll holder, chrome 15.00–19.00
    Towel bar, chrome, 18" 10.00–30.00
    Towel ring, chrome 9.00–13.00
Bathtub and Shower
    Bathtub, white porcelain on steel, 60" x 30" $75.00-200.00
    Bathtub surround kit 40.00-150.00
    Bathtub enclosure 85.00-180.00
    Shower or tub/shower unit 200.00–300.00
    Whirlpool tubs 850.00–1700.00
    Faucets, brass $150.00–190.00
    Faucets, chrome $70.00–125.00
Medicine Cabinet
    Swing door with mirror, 24" x 26" $20.00–40.00
    Oak, 24" x 26" 60.00
Sink
    White china, 20" x 17" oval $38.00
    Pedestal, vitreous china, 32" high 40.00–160.00
    Faucets, brass or chrome 70.00–130.00
Toilet
    Vitreous china, water-saving $55.00–100.00
    Vitreous china, elegant styling 230.00–300.00
Vanity
    24" x 21" $80.00–100.00
    30" x 18" (2-Door, 2-Drawer) 155.00
    30" x 18" (1-Door, 2-Drawer) 140.00

Chapter 4

ROOM CONVERSIONS

Converting spaces such as attics to living space is one of the strongest marketing opportunities for the remodeling contractor. Attic, garage and basement conversions are a multi-billion dollar segment of the remodeling market and have become a national trend.

Before starting a conversion it is important that you check local building codes, because most cities require a building permit for any alteration to a home. Many cities will not give out building permits for remodeling unless services such as plumbing, heating and electrical are brought up to current code standards, even though you did not plan on these types of work. You will also need to plan the number of heating, plumbing and electrical outlets and add them to your estimate.

Insulation is another area in which you should check the local building code. R-19 for walls and R-30 for ceilings have become common standards. And in attic conversions you should recommend sound deadening to your customers. The room may be beautiful, but if all of the sounds from the room penetrate the living space below, you might have an unhappy client.

ATTICS

In many older homes the attic is the ideal place to add living space, because these houses have attics. The remodeler does not need to worry about exterior walls and roofing . However, in newer homes with roof truss construction, such conversions are most likely infeasible.

An attic conversion can be as simple or as elaborate as a homeowner wishes, depending on his pocketbook and the space available. Such things as adding stairways and upgrading heating, electrical and plumbing service are detailed in other chapters. Here, we will discuss what applies specifically to the attic.

Dormers and skylights can be added for light. Louvers and other types of vents or fans should be added for proper ventilation. Some attics already have sufficient head room, but others will require dormers to increase space. In still others the ceiling can be finished with a cathedral effect. The possibilities are unlimited.

The first thing to consider is the flooring. In many homes the ceiling joists are 2" x 6"s, which are inadequate for supporting a floor for a living area. Building codes usually require a live floor design capable of handling 30 lbs. per sq. in.

This problem can usually be remedied in one of two ways. One is to simply double existing joists. The other is to cut and install 2" x 4" solid bridging, 24" o.c. to distribute the load. The blocks should be staggered for easy nailing, as in the illustration below.

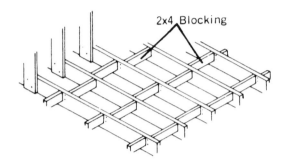

For extra support, add 2 x 4 blocks 24" o.c.

Figure one carpenter can cut and install 24 lin.ft. of blocks per hour at the following cost per lin. ft.:

	Hours	Rate	Total	Rate	Total
Cost per l.f.	0.042	--	--	$25.60	$1.08

The recommended subfloor for attics is 4' x 8' x 3/4" plywood panels. The panels should be laid across the joists, using 8d nails, nailed about every 6" along the outside edge and about every 8" along the joists. The butt end of all panels should meet exactly on the joist; however, the joints should be staggered so that the butt ends of all panels do not meet on the same joist. See Chapter 6, *Stairways and Flooring*.

Installing New Windows. When installing new windows at gable ends, double 2" x 4"s or 2" x 6"s can be stood on an edge for the header, depending on the size of the opening.

For the sill a 2" x 4" laid flat is adequate. Use 16d nails to nail through the studs into the header. Add additional 2" x 4" bracing below the window sill to support the additional weight of the new window. For additional information and estimating, see Chapter 14, *Windows and Doors*.

ROOM CONVERSIONS 37

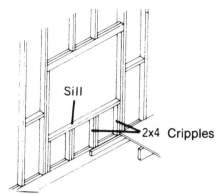

Double 2 x 4's under windowsill for additional support.

Installing Skylights. A popular skylight size is 32" x 32". To cut the opening for this size means cutting only one common rafter. Add braces to the rafters above and below the rafter being cut to support the roof. The shingles above the opening should be carefully removed to avoid damage to the roof and adjacent shingles.

You need to cut a hole in the roof that is the right size to accommodate the skylight you are installing. For a 32" x 32" skylight, frame the opening on the roof using 2" x 6" with the inside dimension of the framing measuring 30-3/8" x 30-3/8". Hold the frame square by nailing 1" x 3" blocks in each corner. Place the frame over the opening and mark a chalk line around the outside perimeter on the roof. Cut all shingles along the chalk line about 1-5/8" back from the edge of the opening. Temporarily loosen all shingles within 4" of the cut opening.

Cut and bend four strips of aluminum flashing to form the flange around the curb frame. For a 32" x 32" skylight, the flashing should be cut 9" x 39" and folded so that it covers the top and side of the frame and extends 3" onto the roof. The corners should be cut and bent as shown. Apply roofing cement or sealant between the flashing, the roof paper and the shingles. A layer of sealant between the flashing and the top of the curb also helps prevent leaks.

Toenail the frame on the inside into the rafters using 8d nails. Nail the flashing to the frame and the roof with 16d common nails. The skylight can now be fastened into position.

Figure one carpenter can complete the preceding framing in 5 hours. To install the skylight itself, add 2 hours.

Installing Walls, Partitions and Ceilings. Before the partitions or ceilings are started, it is customary to frame the walls. Wall height can be 4' or 5' depending on the wishes of the homeowner.

An attic is often the ideal place to add living space, and skylights can fill an attic room with natural light.

A plumb line is used to snap a chalk line from the rafter to the floor at the proper height to determine the position of the sole plate. Use a 2" x 4" for the sole plate, secured to the floor with 16d common nails. Nail 2" x 3" or 2" x 4" studs to each rafter and the sole plate at the desired room height. Check studs with a level to be sure they are plumb in all directions. To provide a nailing surface for the wall material, nail blocking between rafters at the top of the knee wall. If the area behind the wall is to be used for storage, use a 2" x 4" for the header over the door opening that provides access to the storage area.

Ceiling joists can now be nailed to opposite rafters to frame the ceiling. These can be 7'-6" to 8' above the floor.

ROOM CONVERSIONS

If a dividing wall is desired, it can be anchored to the ceiling joist and placed at any position in the attic. Similarly, a partition wall can also be placed in any position. Both are constructed by placing a sole plate in the proper location and framing out the wall by placing studs 16" o.c.

Figure one carpenter can frame 500 lin. ft. of wall and ceiling per day at the following cost per lin. ft.:

	Hours	Rate	Total	Rate	Total
Cost per l.f.	0.016	--	--	$25.60	$0.41

Adding a Dormer. Gabled dormers are the most difficult to install, because they involve the placement of valley rafters and are constructed so that the pitch of the dormer roof is different from the main roof. Dormers with side walls should be positioned so that the side walls are located over roof rafters on both sides. These rafters are doubled to support the exterior studs and valley rafter.

The valley rafters are tied into a header that in turn is tied into the ridge with jack rafters. The dormer is then completed with conventional framing. Care should be taken to provide proper flashing where the dormer walls intersect the roof.

Material Checklist for Gabled Dormer
(Roofing, Siding, Painting and Interior Finishes Not Included)

Materials and Size		5'	6'	8'	10'	12'
Rafter Headers	2" x 6"	2-10'	2-12'	2-16'	4-10'	4-12'
Rafter Plates	2" x 6"	2-16'	2-16'	2-16'	2-16'	2-16'
Studs and Plates						
Side	2" x 4"	2-10'	2-10'	2-10'	2-10'	2-10'
Side	2" x 4"	2-12'	2-12'	2-12'	2-12'	2-12'
Front	2" x 4"	7-10'	8-10'	11-10'	13-10'	11-12'
Gable	2" x 4"	1-12'	1-12'	3-10'	2-10'	2-12'
Gable	2" x 4"	1-8'	1-8'	1-10'	1-14'	1-14'
Rafters	2" x 4"	6-8'	6-10'	6-12'	6-12'	6-14'
Ridge Boards	1" x 6"	1-10'	1-10'	1-10'	1-10'	1-10'
Framing, total b.f.		179	200	237	264	283
Sheathing, b.f.	1" x 6"	60	70	70	85	80
Roofing						
boards, b.f.	1" x 6"	60	72	86	100	120

Cornice						
Fascia, l.f.	1" x 6"	22	24	26	28	30
Crown						
Mould, l.f.	1" x 3"	22	24	26	28	30
Nails, lbs.		10	11	12	13	14
Flashing, l.f.	12" wide	33	35	39	42	46
Felt, rolls	30-lb.	1	1	1	1	2
Window frame	2/6 x 4/10	1	1	2	2	3
Window sash	2/6 x 4/10	1	1	2	2	3
Labor						
to install, hrs.		26	27	32	36	40

Gabled Dormer Roof and Side Areas for 1/2 Pitch Roofs

	Width of Dormer				
	5'	6'	8'	10'	12'
Roof, sq. ft.	60	72	86	100	120
Wall, sq. ft.	60	70	70	85	80
Starter, lin. ft.	10	10	10	10	10
Ridge, lin. ft.	12	12	12	12	12

Shed Dormers. Shed dormers are much simpler to construct and are tied directly to the ridge. Side walls are also finished off using conventional framing procedures. The common rafters supporting the outer walls should be doubled the same as for gabled dormers. Figure two carpenters are required to complete the following:

Dormer Width	Days to Install
10'	4
16'	6
20'	7
24'	8

The above includes:
1. 2" x 4" studs, 16" o.c.
2. 1/2" gypsum sheathing
3. Aluminum siding
4. S.I.S. roofing
5. One window for each 10 lin. ft. of exterior wall

ROOM CONVERSIONS

For additional windows, add $75.00 (includes material and labor). For insulated siding over 1" x 4" wood stripping over gypsum, add $75.00 per square to install furring strips.

To determine siding costs for the wall areas of dormers, use the following table. Window areas have already been deducted:

Sq. Ft. of Side Wall Area for Dormers

Dormer Width	Lift Type	Type of Dormer Gable Type	Hip Type
5'-0" (One Window)	88	60	52
6'-0"	94	70	58
7'-0"	100	80	64
8'-0" (Two Windows)	88	70	52
9'-0"	94	75	58
10'-0"	100	85	64
11'-0"	106	90	70
12'-0" (Three Windows)	93	80	57

Add the net square feet of each elevation, gable and dormer together, add 8% for waste and figure to the nearest larger half or full square or nearest bundle of shingles for material costs.

You might want to strip off the old siding before new is applied, or you might want to cover it over, either by applying an underlayment board or by applying wood stripping and felt over the existing siding. Underlayment board would figure the same quantity as siding. Wood stripping can be figured per square, depending on the exposure of existing siding, as follows:

Exposures in inches	4"	4-1/2"	5"	5-1/2"	6"	6-1/2"	8"
Lin. ft. of stripping	300	280	240	224	200	188	152

For corners, use full height for metal corners. For corner boards, figure one 4" and one 6" board for each corner. Add molding, flashing and caulking costs at doors and windows.

Labor per Square and Clapboards Applied with Wood Stripping*

Size	Type	No. of Pcs. per Sq.	Carpenter Hours
12" x 24"	Shingle	57	2.66
8" x 24"	Shingle	93	3.50
9-1/2" x 8'	Clapboard	19	2.66

*See Chapter 12.

BASEMENTS

When it comes to converting basements into family or recreation rooms, the biggest problem you will probably run into is moisture. Sometimes a dehumidifier can reduce this problem, but a waterproof seal is recommended as the first step.

Several waterproof coatings are available to alleviate moisture problems with the walls. The coverage rates for several different types are given in the tables below. The rates are furnished by the manufacturers. Note that some specify one coat, and others call for two coats. Regardless of the material used, a sufficient quantity must be applied to completely seal the surface. Otherwise, results will be unsatisfactory.

Cost of 100 Sq. Ft. Liquid Water Repellent

These transparent liquids are used for dampproofing brick, stone, stucco, concrete and cement surfaces without changing the color of the surface treated.

	No. Coats	S.F. per Gal.	Gals. per 100 S.F.	Price per Gal.	Labor Hrs. to Apply
Daracone	1	75-100	1 1/4	$11.00	1.6
Supertox	1	125-150	3/4	9.80	0.8
Tremco 141 Invisible	2	75-100	1 1/4	9.25	1.6

Silicone Base Water Repellents. These allow the masonry to "breathe" while rendering the surface water repellent. They penetrate deeply into the cement of masonry surfaces, coating pores, cracks and fissures with an insoluble, non-oxidizing film of silicone that effectively stops the capillary action by which water is absorbed. The covering capacities and quantities in the following table were furnished by the manufacturers.

Cost of 100 Sq. Ft. Silicone Base, Transparent Liquid Water Repellent

Name of Material	No. Coats	S.F. per Gal.	Gals. per 100 S.F.	Labor Hrs. to Apply
Dehydratine 22	1	75-100	1 1/4	0.8
Hydrocide Colorless SX	1	75-100	1	0.8
Sika Transparent	1	80-200	2/3	0.8
Tremco 147-3%	1	100-200	2/3	0.8
Tremco 147-5%	1	100-200	2/3	0.8
Devoe Super-Por-Seal	1	150	2/3	0.8

Labor Applying Transparent Liquid Waterproofing. When applying transparent liquid waterproofing by hand, a worker should be able to cover 900 to 1100 sq. ft. per 8-hr. day at the following labor cost per sq. ft.:

	Hours	Rate	Total	Rate	Total
Roofer	0.008	--	--	$24.60	$0.20

On the second and third coats, a worker should apply 1000 to 1200 sq. ft. per 8-hr. day at the following cost per sq. ft.:

	Hours	Rate	Total	Rate	Total
Roofer	0.007	--	--	$24.60	$0.17

If the concrete slab floor is in good condition and dry, several different types of flooring, such as various tile or carpeting, can be applied directly to the floor by following the manufacturer's recommended application procedure.

If the flooring is damp, then some type of dampproofing should be applied. A good vapor barrier is probably the best solution. One method is to follow the procedure for applying strip flooring over concrete outlined in Chapter 6, *Flooring and Stairways*. Another is to apply a mastic followed by a vapor barrier. Generally, treated plywood panels are laid over the sleepers for the subfloor and the finish flooring is applied according to manufacturer's recommendations.

Placing Wood Floor Sleepers. When placing 2" x 3" or 2" x 4" wood floor screeds or sleepers over concrete floors to receive finish flooring, a carpenter should place from 225 to 275 lin. ft. per 8-hr. day at the following labor cost per lin. ft.:

	Hours	Rate	Total	Rate	Total
Carpenter	0.032	--	--	$25.60	$0.82
Labor.	0.008	--	--	21.50	0.17
Cost per l.f.	0.040				$0.99

For *first-grade workmanship* 2" x 3" or 2" x 4" beveled floor sleepers are placed over the rough concrete floors and wedged or blocked up to provide a perfectly level surface to receive the finish flooring. The screeds are anchored with special concrete nails or fasteners.

A carpenter should be able to place, wedge and level 130 to 170 lin. ft. of sleepers per 8-hr. day at the following cost per lin. ft.:

	Hours	Rate	Total	Rate	Total
Carpenter	0.053	--	--	$25.60	$1.36
Labor.	0.008	--	--	21.50	0.17
Cost per l.f.	0.061				$1.53

Walls. To finish walls, 2" x 2" furring strips are nailed to the wall 16" o.c. These are fastened at the bottom and top to 2" x 2" sole and top plates. If dampproofing was required, the 2" x 2"s should be treated to prevent decay due to moisture using standard masonry nails 24" apart.

Another option is to build a standard wall using 2" x 4" studs. This method provides space for wiring, plumbing and heating. Also, no nails or anchors are put into the walls, because the top plate is fastened to the floor joists and sole plate to the floor.

Placing Wood Furring Strips on Masonry Walls. If the walls are straight and plumb, a carpenter should be able to place 500 to 550 lin. ft. of furring strips per 8-hr. day at the following cost per lin. ft.:

	Hours	Rate	Total	Rate	Total
Cost per l.f.	0.015	--	--	$25.60	$0.38

For *first grade workmanship* where it is necessary to plug the masonry wall and place all furring strips straight and plumb, a carpenter should place 200 to 250 lin. ft. of strips per 8-hr. day at the following cost per lin. ft.:

	Hours	Rate	Total	Rate	Total
Cost per l.f.	0.036	--	--	$25.60	$0.93

Insulation. Fiberglass or Rock Wool batts should be placed with the vapor barrier toward the side of the wall that is warm in winter. Faced insulation can be stapled by "inset" or "face" stapling. A carpenter should be able to insulate 2500 sq. ft. of walls per 8-hr. day at the following labor cost per sq. ft.:

	Hours	Rate	Total	Rate	Total
Cost per s.f.	0.0032	--	--	$20.60	$0.08

The walls are now ready for finishing. See Chapters 7 and 8. The most economical and efficient ceiling are those covered with ceiling tile or suspended ceilings using suspended ceiling grid metal. Not only are they attractive and simple to install, they allow easy access to electrical and plumbing lines if repairs are required. For ceiling treatments, see Chapter 7, *Drywall, Paneling and Ceiling Tile.*

Typical Prices for Framing Material

Keep in mind that these prices are simply national averages meant to give one a general idea of material costs. The contractor must always check with his own material suppliers for accurate pricing. Many new products are coming into use in residential remodeling and construction, and the remodeler should check into the feasibility and cost effectiveness of these materials in his area. We have included below some sample pricing for metal studs, which are a viable alternative to traditional framing materials.

Studs

2" x 4"-8' (economy grade, 92-5/8") .... $1.39 ea.
2" x 4"-8' (top quality, kiln-dried) .... 2.86 ea.
2" x 4"-7' (construction) .... 1.66 ea.

2 x 4 Construction
  10'-16'..$462.00 M
  18'-20'..588.00 M
  22'-24'..714.00 M

2 x 6 Construction
  8'-16'..460.00 M
  18'-20'..588.00 M
  22'-24'..714.00 M

2 x 8 Construction
  8'-16'..504.00 M
  18'-20'..588.00 M
  22'-24'..714.00 M

2 x 10 Construction
  8'-16'..546.00 M
  10'-20'..588.00 M
  22'-24'..714.00 M

2 x 12 Construction
  8'-16'..588.00 M
  18'-20'..630.00 M
  22'-24'..840.00 M

4 x 4 Construction
  8'-20'..588.00 M

Metal Studs, 25 Gauge
  3-5/8" x 8' stud..1.29 ea.
  3-5/8" x 10' stud..1.59
  3-5/8" x 10' runner 1.77

Sanded Trim Boards (Select)

'D' Pine*		'D' Pine*		#1 and #2 Pine*	
1 x 2	$0.32	1/2 x 2	$0.43	5/4 x 2	$0.32
1 x 3	0.47	1/2 x 3	0.64	5/4 x 3	0.48
1 x 4	0.60	1/2 x 4	0.85	5/4 x 4	0.63
1 x 6	0.90	1/2 x 6	1.28	5/4 x 6	0.95
1 x 8	1.20	1/2 x 8	1.70	5/4 x 8	1.25
1 x 10	1.50	1/2 x 10	2.15	5/4 x 10	1.60
1 x 12	1.80	1/2 x 12	2.55	5/4 x 12	1.90

*per lin. ft.

ROOM CONVERSIONS

Plywood

Sanded Plywood
4' x 8' Sheets - Exterior Glue

Good One-Side	Sq. Ft.	Sheet
1/4 AC/BC	$0.50	$16.00
3/8 AC/BC	0.55	17.60
1/2 AC/BC	0.65	20.80
3/4 AC/BC	0.80	25.60

Sheathing Plywood
4' x 8' Sheets - Exterior Glue

Rough-Two Sides	Sq. Ft.	Sheet
3/8 CDX	$0.45	$14.40
1/2 CDX (4-ply pine)	0.55	17.60
5/8 CDX	0.65	20.80
3/4 CDX	0.75	24.00

Hardboard Masonite
4' x 8' Sheets

Standard		Oil Treated	
1/8"	$5.36 ea.	1/8"	$6.71 ea.
1/4"	7.60 ea.	1/4"	8.96 ea.

Chapter 5

ROOM ADDITIONS

One possible option for adding a room is to convert an attached garage to living space. There are two things to check and correct right away. First, most garage floors are sloped to the door to provide for drainage, so when putting in the new floor, the sleepers should be shimmed up so that flooring is level throughout. This procedure is described in the last chapter. Second, usually there are no foundations under the door area. Since this area will probably be framed in and enclosed, it will have to be dug out and a new foundation and footings poured extending to the existing side foundations.

If building codes permit, the contractor can install the electrical service below the flooring and offer the homeowner a savings in labor and materials. Since most slabs are on grade, they should be insulated around the perimeter and underneath the flooring. Insulation should also be installed in the walls and ceiling.

Perimeter Insulation. This can be accomplished by digging out around the foundation down to the depth of the footing. When used as recommended, expanded polystyrene qualifies for slab-on grade perimeter insulation. To meet HUD Minimum Property Standards, the following thicknesses of expanded polystyrene are recommended:

Unheated Slabs		Heated Slabs	
Winter degree days	EPS thickness	Winter degree days	EPS thickness
4000 or less	1"	2000 or less	1"
4001 to 8000	1-1/2"	2001 to 3000	1-1/4"
8001 to more	2"	3001 to 5000	1-1/2"
		5001 to 8000	2"
		8001 or more	2-1/2"

To comply with FHA moisture protection requirements, use a vapor barrier rated at one perm or less. In most cases no attachment other than backfill is required, and a worker should be able to place approximately 3200 sq. ft. of expanded polystyrene board stock per 8-hr. day at the following labor cost per 100 sq. ft.:

	Hours	Rate	Total	Rate	Total
Carpenter	0.25	--	--	$25.60	$6.40
Cost per s.f.					0.06

Labor Insulating Walls and Floors. For most applications, fiber glass or mineral wood batts and blankets are used. They are supplied in three forms: foil-faced, kraft paper-faced and unfaced. Where the materials are faced, the facings act as a vapor barrier. Unfaced material requires a separate vapor barrier. For the actual perm rating of unfaced material, check the manufacturer's literature. On faced material, the facing extends beyond the insulation to form a stapling flange. Unfaced material is friction fit. Both fiber glass and mineral wool are applied in the same manner. Mineral wool, because it is stiffer, takes a bit longer to install. On slab floors, such as garages, the insulation is simply placed between the sleepers with the vapor barrier to the warm side. An experienced carpenter should be able to staple and fit 2000 to 3000 sq. ft. of insulation per 8-hr. day at the following labor cost per 100 sq. ft.:

	Hours	Rate	Total	Rate	Total
Carpenter	0.32	--	--	$25.60	$8.19
Cost per s.f.					0.08

Labor Insulating Ceilings. An experienced carpenter should be able to staple and fit 1000 to 1800 sq. ft. of insulation per 8-hr. day at the following labor cost per sq. ft.:

	Hours	Rate	Total	Rate	Total
Cost per s.f.	0.006	--	--	$25.60	$0.15

The doorway can be enclosed by using regular 2" x 4" studs. Estimating this work is described later in this chapter and in Chapter 14, *Windows and Doors.* To install windows and doors and electrical service, see Chapters 10 and 14.

Room Additions. Room additions are basically new construction. The biggest problem occurs in matching the siding and roofing. Another problem in adding a room to either end of the house is whether there is sufficient property to permit the addition and still meet local zoning regulations. More often than not, this is why new rooms are added at right angles to the back of the house. Placement of a new room, at either end or the center, is frequently dictated by the entry way. Two other things to consider are plumbing and electrical. How far away is the water supply for any new fixtures and how far away is the wastepipe into which it will drain. The same applies to electrical. Is there a box nearby or does a new one have to be installed? By carefully figuring these installations, a lot of headaches and costs can be avoided.

Special attention must be given to matching siding, windows and roofing with the existing structure.

Most room additions are placed on concrete slabs. Whether footings are required depends on local codes. Figure a 20' x 20' slab as follows. Forming below grade for the foundation is usually not necessary. The foundation will require a trench 6" wide by 42" deep. Figure one laborer using a power trencher will dig 45 to 50 lin. ft. per hour at the following cost per lin. ft.:

	Hours	Rate	Total	Rate	Total
Cost per l.f.	0.021	--	--	$21.50	$0.45

In forming for the slab, figure one carpenter can set forms using stakes 3' to 4' apart at approximately 40 lin. ft. per hour:

	Hours	Rate	Total	Rate	Total
Cost per l.f.	0.025	--	--	$25.60	$0.64

For fill below the slab, one laborer should be able to tamp and level 20' x 20' in one hour. To place reinforcing using 6" mesh, add 0.5 hours to the above. Add for dirt removal and trucking, if required.

For finishing reinforced concrete, the average will vary from 70 to 80 sq. ft. per hour or 560 to 640 sq. ft. per hour 8-hour day the following labor cost per 100 sq. ft.:

	Hours	Rate	Total	Rate	Total
Cement Mason	1.3	--	--	$24.90	$32.37
Laborer	1.3	--	--	21.50	27.95
Cost per 100 s.f.					$60.32
Cost per s.f.					0.60

Labor to Frame Walls

Material	Size	b.f. or l.f. per hr.
Partition Studs		50 b.f.
Partition Plates and Shoe		50 b.f.
Wall Backing		50 b.f.
Grounds		85 l.f.
Knee Wall Plates	2"x4"	40 b.f.
	2"x6"	40 b.f.
Knee Wall Studs	2"x4"	40 b.f.
	2"x6"	50 b.f.

ROOM ADDITIONS

Material	Size	b.f. or l.f. per hr.
Outside Studs	2"x4"	40 b.f.
	2"x6"	50 b.f.
Outside Wall	2"x4"	40 b.f.
Plates and Shoe	2"x6"	50 b.f.
Headers	2"x4"	40 b.f.
	2"x6"	50 b.f.
Gable-end Studs		50 b.f.
Fire Stops		50 b.f.
Corner Braces		50 b.f.

Labor to Apply Sheathing

Material	Size	Rate
Wood (horizontal application)	1" x 6"	65 bf/hr
	1" x 8"	70 bf/hr
	1" x 10"	75 bf/hr
Gypsum Board	48" x 96"	100 sf/2.2 hr
Plywood	48" x 96"	100 sf/2.2 hr

Material and Labor to Install Ceiling Joists

Joist Size	\multicolumn{4}{c}{B.f. Required per 100 s.f.}	Nails (lbs. per 1000 b.f.)	Labor b.f./hr.
	12" o.c.	16" o.c.	20" o.c.	24" o.c.		
2" x 4"	78	59	48	42	19	60
2" x 6"	115	88	72	63	13	65
2" x 8"	153	117	96	84	9	65
2" x 10"	194	147	121	104	7	70
2" x 12"	230	176	144	126	6	70

Although it is not included in the chart above, engineered lumber joists are increasing popular in place of 2 x 10 and 2 x 12 framing. Engineered wood products are free of many of the defects in conventional lumber, such as warping.

ESTIMATING LUMBER QUANTITIES

When estimating the quantity of lumber required for any job, the only safe method is to take off every piece of lumber required to complete that portion of the work. The following tables are given to simplify the work as much as possible and at the same time provide accurate material quantities.

Estimating Wood Joists. When estimating wood joists, always allow 4" to 6" on each end of the joist for bearing on the wall. To obtain the number of joists required for any floor, take the length of the floor in feet, multiply the distance the joists are spaced and add 1 to allow for the extra joist required at end of span.

For example, a floor is 28' long and 15' wide. It will require 16' joists to allow for wall bearing at each end. Assuming the joists are spaced 16" on centers, one joist will be required every 16" or every 1-1/3 or 1.334'. This means it requires three-fourths as many joists as number of feet in the span plus one extra joist. An easy way to determine the number of joists required is to divide 12 by the o.c. spacing in inches. In this case, 12 ÷ 16 = 0.75. Now, multiply the floor length by 0.75 and then add one for the extra joist: 28 x 0.75 = 21 + 1 = 22. You need 22 joists for this section. The following table gives the number of joists required for any spacing:

Number of Wood Floor Joists Required for any Spacing

Distance Joists are Placed on Centers	Multiply Length of Floor Span by	Add Joists	Distance Joists are Placed on Centers	Multiply Length of Floor Span by	Add Joists
12"	1	1	36"	1/3 or .33	1
16"	3/4 or .75	1	42"	2/7 or .29	1
20"	3/5 or .60	1	48"	1/4 or .25	1
24"	1/2 or .50	1	54"	2/9 or .22	1
30"	2/5 or .40	1	60"	1/5 or .20	1

ROOM ADDITIONS

Number of B.F. of Lumber Required per 100 Sq. Ft. of Surface When Used for Studs, Joists, Rafters, Wall and Floor Furring Strips

The following table does not include any allowance for waste in cutting, for doubling joists under partitions or around stairwells or for extra joists at end of each span, top or bottom plates. These items vary with each job. Add as required.

Size of Lumber	12-inch Centers	16-inch Centers	20-inch Centers	24-inch Centers
1" x 2"	16-2/3	12-1/2	10	8-1/3
2" x 2"	33-1/3	25	20	16-2/3
2" x 4"	66-2/3	50	40	33-1/3
2" x 5"	83-1/3	62-1/2	50	41-2/3
2" x 6"	100	75	60	50
2" x 8"	133-1/3	100	80	66-2/3
2" x 10"	166-2/3	125	100	83-1/3
2" x 12"	200	150	120	100
2" x 14"	233-1/3	175	140	116-2/3
3" x 6"	150	112-1/2	90	75
3" x 8"	200	133-1/3	120	100
3" x 10"	250	187-1/2	150	125
3" x 12"	300	225	180	150
3" x 14"	350	262-1/2	210	175

Number of Wood Joists Required for any Floor and Spacing

Length of Floor	Spacing of Joists									
	12"	16"	20"	24"	30"	36"	42"	48"	54"	60"
6	7	6	5	4	3	3	3	3	2	2
7	8	6	5	5	4	4	3	3	3	2
8	9	7	6	5	4	4	3	3	3	3
9	10	8	6	6	5	4	4	3	3	3
10	11	9	7	6	5	4	4	4	3	3
11	12	9	8	7	5	5	4	4	3	3
12	13	10	8	7	6	5	4	4	4	3
13	14	11	9	8	6	5	5	4	4	4
14	15	12	9	8	7	6	5	5	4	4

Number of Wood Joists Required for any Floor and Spacing

Length of Floor	\multicolumn{10}{c}{Spacing of Joists}
	12"	16"	20"	24"	30"	36"	42"	48"	54"	60"
15	16	12	10	9	7	6	5	5	4	4
16	17	13	11	9	7	6	6	5	5	4
17	18	14	11	10	8	7	6	5	5	4
18	19	15	12	10	8	7	6	6	5	4
19	20	15	12	11	9	7	6	6	5	5
20	21	16	13	11	9	8	7	6	5	5
21	22	17	14	12	9	8	7	6	6	5
22	23	18	14	12	10	8	7	7	6	5
23	24	18	15	13	10	9	8	7	6	6
24	25	19	15	13	11	9	8	7	6	6
25	26	20	16	14	11	9	8	7	7	6
26	27	21	17	14	11	10	8	8	7	6
27	28	21	17	15	12	10	9	8	7	6
28	29	22	18	15	12	10	9	8	7	7
29	30	23	18	16	13	11	9	8	7	7
30	31	24	19	16	13	11	10	9	8	7
31	32	24	20	17	13	11	10	9	8	7
32	33	25	20	17	14	12	10	9	8	7
33	34	26	21	18	14	12	10	9	8	8
34	35	27	21	18	15	12	11	10	9	8
35	36	27	22	19	15	13	11	10	9	8
36	37	28	23	19	15	13	11	10	9	8
37	38	29	23	20	16	13	12	10	9	8
38	39	30	24	20	16	14	12	11	9	9
39	40	30	24	21	17	14	12	11	10	9
40	41	31	25	21	17	14	12	11	10	9

One joist has been added to each of the above quantities to account for an extra joist at the end of a span. Add for doubling joists under all partitions.

Estimating Quantity of Bridging. It is customary to place a double row of bridging between joists about 6'-0" to 8'-0" on centers. Joists that are 10'-0" to 12'-0" long require one double row of bridging or 2 pieces to each joists. Joists 14'-0" to 20'-0" long require two double rows of bridging or 4 pieces to each joist. Bridging is usually cut from 1" x 3", 1" x 4", 2" x 2" or 2" x 4" lumber.

ROOM ADDITIONS

The following table gives the approximate number of pieces and the lin. ft. of bridging required per 100 sq. ft. of floor.

Joists up to 12 Feet Long				Joists up to 20 Feet Long			
12" O.C.		16" O.C.		12" O.C.		16" O.C.	
Pcs.	Lin.Ft.	Pcs.	Lin. Ft.	Pcs.	Lin.Ft.	Pcs.	Lin.Ft.
20	30	16	24	40	60	32	48

Estimating Number of Wood Studs. When estimating the number of wood partition studs, take the length of each partition and the total length of all partitions.

If a top and bottom plate is required, take the length of the wood partition and multiply by 2. The result will be the number of lin. ft. of plates required.

If a double plate consisting of 2 top members and a single bottom plate is used, multiply the length of the wood partitions by 3.

For example, find the quantity of lumber required to build a stud partition 16'-0" long and 8'-0" high, with studs spaced 16" on centers and having single top and bottom plates: 16'-0" = 192"; 192" ÷ 16" = 12 studs, plus one extra at the end = 13 8-foot studs. Top and bottom plates, each 16'-0" long, equals 32 lin. ft.

```
13 pcs. 2" x 4" x 8' =    104
13 pcs. 2" x 4" x 16' =    32
Total                     136
136 l.f. x 2/3 = 90.67 board feet (b.f.)
```

After the total number of lin. ft. of lumber has been obtained, convert to board measure.

Board Feet of Lumber Required for Wood Stud Partitions
(2" x 4" studs, 16" on centers, with single top and bottom plates)

Length of Partition	Height of Partition				
	8'-0"	8'-6"	9'-0"	10'-0"	12'-0"
3'-0"	20	22	22	24	28
4'-0"	27	29	29	32	37
5'-0"	33	37	37	40	47
6'-0"	40	44	44	48	56
7'-0"	41	45	45	49	57
8'-0"	48	53	53	57	67

Length of Partition	Height of Partition				
	8'-0"	8'-6"	9'-0"	10'-0"	12'-0"
9'-0"	55	60	60	65	76
10'-0"	61	67	67	73	85
11'-0"	63	69	69	75	87
12'-0"	69	76	76	83	96
13'-0"	76	83	83	91	105
14'-0"	83	91	91	99	115
15'-0"	84	92	92	100	116
16'-0"	91	99	99	108	125
17'-0"	97	107	107	116	135
18'-0"	104	114	114	124	144
19'-0"	105	115	115	125	145
20'-0"	112	123	123	133	155
21'-0"	119	130	130	141	164
22'-0"	125	137	137	149	173
23'-0"	127	139	139	151	175
24'-0"	133	146	146	159	184
25'-0"	140	153	153	167	193
26'-0"	147	161	161	175	203
27'-0"	148	162	162	176	204
28'-0"	155	169	169	184	213
29'-0"	161	177	177	192	223
30'-0"	168	184	184	200	232
31'-0"	169	185	185	201	233
32'-0"	176	193	193	209	243
33'-0"	183	200	200	217	252
34'-0"	189	207	207	225	261
35'-0"	191	209	209	227	263
36'-0"	197	216	216	235	272
37'-0"	204	223	223	243	281
38'-0"	211	231	231	251	291
39'-0"	212	232	232	252	292
40'-0"	219	239	239	260	301

Add 2/3 b.f. of lumber for each lin. ft. of double top or bottom plate.

This sunroom addition provides extra living space and serves as a passive solar heat source. This unit incorporates Pella aluminum-clad casement windows

ROOM ADDITIONS

for ventilation and contemporary French sliding door for outdoor access. Built-in narrow-slat blinds block sun.

Number of Partition Studs Required for Any Spacing

Distance O.C. Spacing	Multiply Length of Partition by	Add Wood Studs
12"	1.00	1
16"	0.75	1
20"	0.60	1
24"	0.50	1

Add for top and bottom plates.

Number of Feet of Lumber Required per Sq. Ft. of Wood Stud Partition Using 2" x 4" Studs

(Studs spaced 16" on centers, with single top and bottom plates)

Partition Length, l.f.	No. of Studs Req'd	Ceiling Height in Feet			
		8'-0"	9'-0"	10'-0"	12'-0"
2	3	1.25	1.167	1.13	1.13
3	3	0.833	0.812	0.80	0.80
4	4	0.833	0.812	0.80	0.80
5	5	0.833	0.812	0.80	0.80
6	6	0.833	0.812	0.80	0.80
7	6	0.833	0.75	0.75	0.80
8	7	0.75	0.75	0.75	0.70
9	8	0.75	0.75	0.75	0.70
10	9	0.75	0.75	0.75	0.70
11	9	0.75	0.70	0.70	0.67
12	10	0.75	0.70	0.70	0.67
13	11	0.75	0.70	0.70	0.67
14	12	0.75	0.70	0.70	0.67
15	12	0.70	0.70	0.70	0.67
16	13	0.70	0.70	0.70	0.67
17	14	0.70	0.70	0.70	0.67
18	15	0.70	0.70	0.67	0.67
19	15	0.70	0.70	0.67	0.67
20	16	0.70	0.70	0.67	0.67
Double plate, add per sf.		0.13	0.11	0.10	0.083

For 2" x 8" studs, double above quantities. For 2" x 6" studs, increase above quantities by 50%.

Example: Find the number of board feet of lumber required for a stud partition 18'-0" long and 9'-0" high. This partition would contain 18 x 9 = 162 sq. ft. The table gives 0.70 b.f. of lumber per sq. ft. of partition. 162 x 0.70 = 113.4 b.f.

Quantity of Plain End Softwood Flooring Required per 100 Sq. Ft. of Floor

Measured Size Inches	Actual Size Inches	Add for Width	B.F. Required per 100 S.F. Surface	Weight per 1000 ft.
1 x 3	3/4 x 2-3/8	27%	132	1800
1 x 4	3/4 x 3-1/4	23%	128	1900

The above quantities include 5% for end cutting and waste.

Quantity of End Matched Softwood Flooring Required per 100 Sq. Ft. of Floor

Measured Size Inches	Actual Size Inches	Add for Width	B.F. Required per 100 S.F. Surface	Weight per 1000 L.F.
1 x 3	13/16 x 2-3/8	27%	130	1800
1 x 4	13/16 x 3-1/4	23%	126	1900

The above quantities include 3% for end cutting and waste.

Quantity of Square Edged (S4S) Boards Required per 100 Sq. Ft. of Surface

Measured Size Inches	Actual Size Inches	Add for Width	B.F. Required per 100 S.F. Surface	Weight per 1000 ft.
1 x 4	3/4 x 3-1/2	14%	119	2300
1 x 6	3/4 x 5-1/2	9%	114	2300
1 x 8	3/4 x 7-1/4	10%	115	2300
1 x 10	3/4 x 9-1/4	8%	113	2300
1 x 12	3/4 x 11-1/4	7%	112	2400

The above quantities include 5% for end cutting and waste.

ROOM ADDITIONS

Lineal Foot Table of Board Measure
Number of Board Feet of Lumber per Lineal Foot

2" x 4" = 0.667	4" x 4" = 1.333	8" x 14" = 9.333
2" x 6" = 1.000	4" x 6" = 2.000	8" x 16" = 10.667
2" x 8" = 1.333	4" x 8" = 2.667	10" x 10" = 8.333
2" x 10" = 1.667	4" x 10" = 3.333	10" x 12" = 10.000
2" x 12" = 2.000	4" x 12" = 4.000	10" x 14" = 11.667
2" x 14" = 2.333	4" x 14" = 4.667	10" x 16" = 13.333
2" x 16" = 2.667	4" x 16" = 5.333	10" x 18" = 15.000
2-1/2" x 12" = 2.500	6" x 6" = 3.000	12" x 12" = 12.000
2-1/2" x 14" = 2.917	6" x 8" = 4.000	12" x 14" = 14.000
2-1/2" x 16" = 3.333	6" x 10" = 5.000	12" x 16" = 16.000
3" x 6" = 1.500	6" x 12" = 6.000	12" x 18" = 18.000
3" x 8" = 2.000	6" x 14" = 7.000	14" x 14" = 16.333
3" x 10" = 2.500	6" x 16" = 8.000	14" x 16" = 18.667
3" x 12" = 3.000	8" x 8" = 5.333	14" x 18" = 21.000
3" x 14" = 3.500	8" x 10" = 6.667	16" x 16" = 21.333
3" x 16" = 4.000	8" x 12" = 8.000	16" x 18" = 24.000

Lengths of Common, Hip and Valley Rafters
per 12 Inches of Run

1	2	3	4	5*	6†
Pitch of Roof	Rise and Run or Cut	Length in Inches Common Rafter per 12" of Run	Percent Common Run	Increase in Rafter over Run	Length in Inches Hip or Valley Rafters
1/12	2 and 12	12.165	0.014	1.014	17.088
1/8	3 and 12	12.369	0.031	1.031	17.233
1/6	4 and 12	12.649	0.054	1.054	17.433
5/24	5 and 12	13.000	0.083	1.083	17.692
1/4	6 and 12	13.417	0.118	1.118	18.000
7/24	7 and 12	13.892	0.158	1.158	18.358
1/3	8 and 12	14.422	0.202	1.202	18.762
3/8	9 and 12	15.000	0.250	1.250	19.209
5/12	10 and 12	15.620	0.302	1.302	19.698

Lengths of Common, Hip and Valley Rafters per 12 Inches of Run

1	2	3	4	5*	6†
11/24	11 and 12	16.279	0.357	1.357	20.224
1/2	12 and 12	16.971	0.413	1.413	20.785
13/24	13 and 12	17.692	0.474	1.474	21.378
7/12	14 and 12	18.439	0.537	1.537	22.000
5/8	15 and 12	19.210	0.601	1.601	22.649
2/3	16 and 12	20.000	0.667	1.667	23.324
17/24	17 and 12	20.809	0.734	1.734	24.021
3/4	18 and 12	21.633	0.803	1.803	24.739
19/24	19 and 12	22.500	0.875	1.875	25.475
5/6	20 and 12	23.375	0.948	1.948	26.230
7/8	21 and 12	24.125	1.010	2.010	27.000
11/24	22 and 12	25.000	1.083	2.083	27.785
11/12	23 and 12	26.000	1.167	2.167	28.583
Full	24 and 12	26.875	1.240	2.240	29.394

*Use figures in this column to obtain area of roof for any pitch. See explanation below.

†Figures in last column are length of hip and valley rafters in inches for each 12 inches of common rafter run.

Area for Roof of Any Pitch. To obtain the number of square feet of roof area for a roof of any pitch, take the entire flat or horizontal area of the roof and multiply by the figure given in the fifth column (*), and the result will be the area of the roof. Always bear in mind that the width of any overhanging cornice must be added to the building area to obtain the total area to be covered. For example, find the area of a roof 26'-0" x 42'-0", having 12" or 1'-0" overhanging cornice. The roof has a 1/4 pitch. To obtain roof area, 26'-0" + 1'-0" + 1'-0" = 28'-0" width. 42'-0" + 1'-0" 1'-0" = 44'-0" length. 28 x 44 = 1,232 sq. ft. of flat or horizontal area.

To obtain area at 1/4 pitch, 1,232 x 1.12* = 1379.84 or 1,380 sq. ft. roof surface. Add allowance for overhang on dormer roofs and sides.

*See column 5 above.

Table of Board Measure
Giving Contents in Feet of Joists, Scantlings and Timbers

Size, inches	\multicolumn{11}{c}{Length in Feet}

Size, inches	10	12	14	16	18	20	22	24	26	28	30
1 x 2	1-2/3	2	2-1/3	2-2/3	3	3-1/3	--	--	--	--	--
1 x 3	2-1/2	3	3-1/2	4	4-1/2	5	--	--	--	--	--
1 x 4	3-1/3	4	4-2/3	5-1/3	6	6-2/3	--	--	--	--	--
1 x 6	5	6	7	8	9	10	--	--	--	--	--
1 x 8	6-2/3	8	9-1/3	10-2/3	12	13-1/3	--	--	--	--	--
1 x 10	8-1/3	10	11-2/3	13-1/3	15	16-2/3	--	--	--	--	--
1 x 12	10	12	14	16	18	20	--	--	--	--	--
1-1/4 x 4	4-1/6	5	5-5/6	6-2/3	7-1/2	8-1/3	--	--	--	--	--
1-1/4 x 6	6-1/4	7-1/2	8-3/4	10	11-1/4	12-1/2	--	--	--	--	--
1-1/4 x 8	8-1/3	10	11-2/3	13-1/3	15	16-2/3	--	--	--	--	--
1-1/4 x 10	10-5/12	12-1/2	14-7/12	16-2/3	18-3/4	20-5/6	--	--	--	--	--
1-1/4 x 12	12-1/2	15	17-1/2	20	22-1/2	25	--	--	--	--	--
1-1/2 x 4	5	6	7	8	9	10	--	--	--	--	--
1-1/2 x 6	7-1/2	9	10-1/2	12	13-1/2	15	--	--	--	--	--
1-1/2 x 8	10	12	14	16	18	20	--	--	--	--	--
1-1/2 x 10	12-1/2	15	17-1/2	20	22-1/2	25	--	--	--	--	--
1-1/2 x 12	15	18	21	24	27	30	--	--	--	--	--
2 x 2	3-1/3	4	4-2/3	5-1/3	6	6-2/3	--	--	--	--	--
2 x 3	5	6	7	8	9	10	11	12	13	14	15
2 x 4	6-2/3	8	9-1/3	10-2/3	12	13-1/3	14-2/3	16	17-1/3	18-2/3	20
2 x 6	10	12	14	16	18	20	22	24	26	28	30
2 x 8	13-1/3	16	18-2/3	21-1/3	24	26-2/3	29-1/3	32	34-2/3	37-1/3	40
2 x 10	16-2/3	20	23-1/3	26-2/3	30	33-1/3	36-2/3	40	43-1/3	46-2/3	50
2 x 12	20	24	28	32	36	40	44	48	52	56	60
2 x 14	23-1/3	28	32-2/3	37-1/3	42	46-2/3	51-1/3	56	60-2/3	65-1/3	70
3 x 4	10	12	14	16	18	20	22	24	26	28	30
3 x 6	15	18	21	24	27	30	33	36	39	42	65
3 x 8	20	24	28	32	36	40	44	48	52	56	60
3 x 10	25	30	35	40	45	50	55	60	65	70	75
3 x 12	30	36	42	48	54	60	66	72	78	84	90
3 x 14	35	42	49	56	63	70	77	84	91	98	105

Table of Board Measure
Giving Contents in Feet of Joists, Scantlings and Timbers

Size, inches	\multicolumn{11}{c}{Length in Feet}
	10	12	14	16	18	20	22	24	26	28	30
4 x 4	13	16	19	21	24	27	29	32	35	37	40
4 x 6	20	24	28	32	36	40	44	48	52	56	60
4 x 8	27	32	37	43	48	53	59	64	69	75	80
4 x 10	33	40	47	53	60	67	73	80	87	93	100
4 x 12	40	48	56	64	72	80	88	96	104	112	120
4 x 14	47	56	65	75	84	93	103	112	121	131	140
6 x 6	30	36	42	48	54	60	66	72	78	84	90
6 x 8	40	48	56	64	72	80	88	96	104	112	120
6 x 10	50	60	70	80	90	100	110	120	130	140	150
6 x 12	60	72	84	96	108	120	132	144	156	168	180
6 x 14	70	84	98	112	126	140	154	168	182	196	210
6 x 16	80	96	112	128	144	160	176	192	208	224	240
8 x 8	53	64	75	85	96	107	117	128	139	149	160
8 x 10	67	80	93	107	120	133	147	160	173	187	200
8 x 12	80	96	112	128	144	160	176	192	208	224	240
8 x 14	93	112	131	149	168	187	205	224	243	261	280
8 x 16	107	128	149	171	192	213	235	256	277	298	320
10 x 10	83	100	117	133	150	167	183	200	217	233	250
10 x 12	100	120	140	160	180	200	220	240	260	280	300
10 x 14	117	140	163	187	210	233	257	280	303	327	350
10 x 16	133	160	187	218	240	267	293	320	347	373	400
12 x 12	120	144	168	192	216	240	264	288	312	336	360
12 x 14	140	168	196	224	252	280	308	336	364	392	420
12 x 16	160	192	224	256	288	320	352	384	416	448	480
14 x 14	163	196	229	261	294	327	359	392	425	457	490
14 x 16	187	224	261	299	336	373	411	448	485	523	560
14 x 18	210	252	294	336	378	420	462	504	546	588	630
14 x 20	233	280	327	373	420	467	513	560	607	653	700

Table of Board Measure
Giving Contents in Feet of Joists, Scantlings and Timbers

Size, inches	Length in Feet										
	10	12	14	16	18	20	22	24	26	28	30
16 x 16	213	256	299	341	384	427	469	512	555	597	640
16 x 18	240	288	336	384	432	480	528	576	624	672	720
16 x 20	267	320	373	425	480	533	587	640	693	747	800
18 x 18	270	324	378	432	486	540	594	648	702	756	810
18 x 20	300	360	420	480	540	600	660	720	780	840	900
20 x 20	333	400	467	533	600	667	733	800	867	933	1000

NAILS REQUIRED FOR CARPENTER WORK

The following table gives the number and lbs. of wire nails required for various work items—per 1,000 b.f. for lumber, per 1,000 shingles or per square (100 sq. ft.) of shingles. A waste percentage is added for loss of material on account of lap or matching of shiplap, flooring, ceiling and siding of various widths. The table gives the size generally used for certain purposes with the nailing space 16" on centers, and 1 or 2 nails per board for each nailing space.

Description of Material	Unit of Measure	Size and Kind of Nail	Nails Required	Lbs of Nails Required
Wood Shingles	1,000'	3d Common	2,560	4
Individual Asphalt Shingles	100 sq. ft.	7/8" Roofing	848	4
Three in One Asphalt Shingles	100 sq. ft.	7/8" Roofing	320	1
Wood Lath	1,000'	3d Fine	4,000	6
Wood Lath	1,000'	2d Fine	4,000	4
Bevel or Lap Siding, 1/2" x 4"	1,000'	6d Coated	2,250	*15
Bevel or Lap Siding, 1/2" x 6"	1,000'	6d Coated	1,500	*10
Byrkit Lath, 1" x 6"	1,000'	6d Common	2,400	15
Drop Siding, 1" x 6"	1,000'	8d Common	3,000	25
3/8" Hardwood Flooring	1,000'	4d Common	9,300	16
25/32" Hardwood Flooring	1,000'	8d Casing	9,300	64
Subflooring, 1" x 3"	1,000'	8d Casing	3,350	23
Subflooring, 1" x 4"	1,000'	8d Casing	2,500	17
Subflooring, 1" x 6"	1,000'	8d Casing	2,600	18
Ceiling, 5/8" x 4"	1,000'	6d Casing	2,250	10
Sheathing Boards, 1" x 4"	1,000'	8d Common	4,500	40
Sheathing Boards, 1" x 6"	1,000'	8d Common	3,000	25
Sheathing Boards, 1" x 8"	1,000'	8d Common	2,250	20
Sheathing Boards, 1" x 10"	1,000'	8d Common	1,800	15
Sheathing Boards, 1" x 12"	1,000'	8d Common	1,500	12.5
Studding, 2" x 4"	1,000'	16d Common	500	10
Joist, 2" x 6"	1,000'	16d Common	332	7
Joist, 2" x 8"	1,000'	16d Common	252	5
Joist, 2" x 10"	1,000'	16d Common	200	4
Joist, 2" x 12"	1,000'	16d Common	168	3.5
Interior Trim, 5/8" thick	1,000'	6d Finish	2,250	7
Interior Trim, 3/4" thick	1,000'	8d Finish	3,000	14
5/8" Trim where nailed to jamb	1,000'	4d Finish	2,250	3
1" x 2" Furring or Bridging	1,000'	6d Common	2,400	15
1" x 1" Grounds	1,000'	6d Common	4,800	30

*Note - Cement coated nails sold as 2/3 lb. = 1 lb. common nails.

ROOM ADDITIONS

Recommended Nailing Schedule for Common Applications in Building Construction

Application		Nail Size Inches	Nail Type	Head Diameter Type		Point Size	Point Type	Nail-ing	Spacing o. c.	Nails Per Joint
	Nailed into									
Mudsill, partition plate, 2"	Concrete	2 1/2–2 3/4×0.148	Sc-1z	5/16"	Checkered	Long	Dia.	Face	12"–24"	—
Ditto, in earthquake regions	Concrete	(3 1/4–) 3 1/2×0.250	Sc-1z	9/16"	Checkered	Med.	Ndl.	Face	24"–48"	—
Ditto, 3"	Concrete	4 1/2×0.250	Sc-1z	9/16"	Checkered	Med.	Ndl.	Face	24"–48"	—
Furring strips	Concrete	1 1/2–1 3/4×0.148	Sc-1z	5/16"	Checkered	Long	Dia.	Face	12"–24"	—
Mudsill	Mudsill	2 1/2×0.120	Sc-2	9/32"	Flat	Med.	Dia.	Toe	—	2
Sleepers	Mudsill	2 1/2×0.120	Sc-2	9/32"	Flat	Med.	Dia.	Toe	—	2
Joists	Mudsill	3 1/4×0.135	Sc-2	5/16"	Flat	Med.	Dia.	Toe	—	2–3
Subflrg., 1" lumber, plywood	Mudsill, sleeper, joist	2 1/8×0.105	St-14	1/4"	Flat, Csk.	Med.	Dia.	Face	6" & 12"	2(3)
Subflrg., 2" lumber, plywood	Mudsill, sleeper, joist	2 7/8×0.120	St-14	9/32"	Flat, Csk.	Med.	Dia.	Face	—	2(3)
Subflrg., 3/8"–1/2" plywood (dph.)	Mudsill, sleeper, joist	1 1/2×0.135	Hi-28	5/16"	Flat, Csk.	Med.	Dia.	Face	6" & 12"	—
Subflrg., 5/8" plywood (dph.)	Mudsill, sleeper, joist	1 3/4×0.135	Hi-28	5/16"	Flat, Csk.	Med.	Dia.	Face	6" & 12"	—
Subflrg., 3/4" plywd. (dph.), part bd.	Mudsill, sleeper, joist	2 ×0.148	Hi-28	5/16"	Flat, Csk.	Med.	Dia.	Face	6" & 12"	—
Subflrg., 1"–1 1/8" plywood (dph.) part. bd.	Mudsill, sleeper, joist	2 1/2×0.148	Hi-28	5/16"	Flat, Csk.	Med.	Dia.	Face	6" & 12"	—
Underlayment, 1/4"–5/16" plywood	Subfloor	1 ×0.083	St-16	3/16"	Flat, Csk.	Med.	Dia.	Face	6" & 6"–12"	—
Underlayment, 3/8"–1/2" plywood	Subfloor	1 1/4×0.083	St-16	3/16"	Flat, Csk.	Med.	Dia.	Face	6" & 6"–12"	—
Underlayment, 5/8" plywood	Subfloor	1 3/8×0.098	St-16	1/4"	Flat, Csk.	Med.	Dia.	Face	6" & 6"–12"	—
Underlayment, 3/4" plywood	Subfloor	1 1/2×0.098	St-16	1/4"	Flat, Csk.	Med.	Dia.	Face	6" & 6"–12"	—
Underlayment, 7/8" plywood	Subfloor	1 5/8×0.098	St-15	1/4"	Flat, Csk.	Med.	Dia.	Face	6" & 6"–12"	—
Underlayment, 3/16"–5/8" hardboard	Subfloor	1–1 3/8×0.083	St-15	3/16"	Flat, Csk.	Med.	Dia.	Face	6" & 6"–12"	—
Flooring, T & G hardwood	Subfloor, joist, sleeper	2–2 1/2×0.115	Sc-4	13/64"	Casing	Blunt	Dia.	Toe	10"–18"	—
Flooring, T & G softwood	Subfloor, joist, sleeper	2–2 1/2×0.115	Sc-4	13/64"	Casing	Blunt	Dia.	Toe	10"–18"	—
Flooring, T & G hardwood, 3/8" and 1/2"	Subfloor, joist, sleeper	1–1 1/4×0.072	Sc-4	9/64"	Casing	Blunt	Dia.	Toe	10"–18"	—
Flooring, T & G parquet	Subfloor	1 1/2×0.105	Sc-4	9/64"	Casing	Blunt	Dia.	Face	—	—
Framing plates	Stud	3 1/4×0.135	Sc-2	5/16"	Flat	Med.	Dia.	Face	16"–24"	2
Framing studs	Stud, cripple, lintel, sill	2 1/2×0.120	Sc-2	9/32"	Flat	Med.	Dia.	Face	—	—
Framing studs	Plate, cripple, lintel, sill	2 1/2×0.120	Sc-2	9/32"	Flat	Med.	Dia.	Toe	16"	3
Framing sole plate	Mudsill	3 1/4×0.135	Sc-2	5/16"	Flat	Med.	Dia.	Face	16"	—
Framing top plate	Lower top plate	3 1/4×0.135	Sc-2	5/16"	Flat	Med.	Dia.	Face	24"	—
Trussed rafter assembly		3 1/4×0.135	Sc-5	5/16"	Flat	Med.	Dia.	Face	2 1/2"–3"	Given
Trussed rafter assembly		2 1/2×0.120	Sc-5	9/32"	Flat	Med.	Dia.	Face	2 1/2"–3"	Given

See Notes, Key to Nail Types and Abbreviations on later page.

Application		Nail Size Inches	Nail Type	Head Diameter	Head Type	Point Size	Point Type	Nailing	Spacing o.c.	Nails Per Joint
	Nailed Into									
Rafter, 4"	Top plate	3 1/4x0.135	Sc-2	5/16"	Flat	Med.	Dia.	Toe	—	3
Rafter, 4"	Top plate	6 x0.177	St-34	7/16"	Flat	Med.	Dia.	Face	—	2
Rafter, 4"	Top plate	7 x0.207	St-34	1/2"	Flat	Med.	Dia.	Face	—	—
Rafter, 6", 8", 10"	Top plate	4-6 x0.177 / 7-9 x0.203	St-34	7/16"	Flat	Med.	Dia.	Toe	—	2-3
Sheathing, 1" lumber	Framing, rafter	2 x0.120	St-3	9/32"	Flat	Med.	Dia.	Face	6" & 12"	2
Sheathing, 3/8"-1/2" plywood	Framing, rafter	13/4x0.120	St-17	9/32"	Flat	Med.	Dia.	Face	6" & 12"	—
Sheathing, 5/16"-1/2" plywood (dph.)	Framing, rafter	11/2x0.135	Hi-17	5/16"	Flat, Csk.	Med.	Dia.	Face	6" & 12"	—
Sheathing, 5/8" plywood (dph)	Framing, rafter	13/4x0.135	Hi-17	5/16"	Flat, Csk.	Med.	Dia.	Face	6" & 12"	—
Sheathing, 3/4" plywood (dph.)	Framing, rafter	2 x0.148	Hi-17	5/16"	Flat, Csk.	Med.	Dia.	Face	6" & 12"	—
Sheathing, 1"-11/8" plywood (dph.)	Framing, rafter	21/2x0.148	Hi-17	5/16"	Flat, Csk.	Med.	Dia.	Face	6" & 12"	—
Sheathing, insulation board, gypsumboard	Framing, rafter	11/2-2 x0.120	Sc-10g	3/8", 7/16"	Flat	Blunt	Dia.	Face	3-4" & 6-8"	—
Sheathing, asbestosboard, 1/8"	Framing, rafter	11/4x0.083	Sc-6g	3/16"	Flat, Csk.	Blunt	Dia.	Face	3-4" & 6-8"	—
Sheathing, asbestosboard, 1/4"	Framing, rafter	11/4x0.120	St-or Sc-6g	5/16"	Flat, Csk.	Blunt	Dia.	Face	3-4" & 6-8"	—
Sheathing, hardboard, 3/8"-5/8"	Framing, rafter	2 x0.115	Sc-7g	13/64"	Flat, Csk.	Med.	Ndl.	Face	3-4" & 6-8"	—
Building paper	Sheathing	1/2-2-3/4x0.105	Sq-30	15/16"	Square	Med.	Dia.	Face	6"-12"	—
Stripping, 3/8"x3 5/8"	Framing, joist, rafter	2 x0.120	St-3	9/32"	Flat	Med.	Dia.	Face	—	2
Stripping, 1"x4"	Framing, joist, rafter	21/2x0.135	St-3	5/16"	Flat	Med.	Dia.	Face	—	2
Stripping, 2"x3"	Framing, joist, rafter	31/2x0.165	St-3	5/8"	Flat	Med.	Dia.	Face	—	2
Siding, wood, 1"	Sheathing and framing	21/8x0.101-0.115	St-14g	1/4"	Flat, Csk.	Med.	Dia.	Face	—	1
Siding, wood, 1"	Sheathing and framing	2 x0.120	Dr-14	5/32"	Flat, Csk.	Med.	Ndl.	Face	—	1
Siding, wood, 2"	Sheathing and framing	3 x0.135	Dr-14	5/32"	Flat, Csk.	Med.	Ndl.	Face	—	1
Siding, plywood	Sheathing and framing	17/8x0.109	Dr-8	5/32"	Casing	Med.	Ndl.	Face	—	—
Siding, T & G wood	Sheathing and framing	13/4x0.105	Sc-8	5/32"	Casing	Med.	Dia.	Toe	6" & 12"	Given
Siding, asbestos shingle	Sheathing and framing	11/2-13/4x0.105	Dr-33	3/16"	Flat Button	Med.	Dia.	Face	6" & 12"	Given
Siding, asbestos shingle	Sheathing and framing	11/2-13/4x0.083	St-19t	3/16"	Flat	Med.	Dia.	Face	—	Given
Siding, asbestos shingle	Sheathing and framing	11/2-13/4x0.076	St-20	3/16"	Flat	Med.	Dia.	Face	8"-12"	2
Siding, insulated brick, wood shingle	Sheathing and framing	13/4x0.095	St-18ge	5/32"	Finishing	Blunt	Dia.	Face	—	2
Siding, wood shingle	Insulating sheathing	13/4-2 x0.083	St-18ge	5/32"	Finishing	Blunt	Dia.	Face	—	2
Siding, wood shingle	Insulating sheathing	13/4-2 x0.105	Dr-18	3/16"	Flat	Long	Dia.	Face	12"	2
Siding, wood shingle	Plywood	11/8x0.102	Dr-18	3/16"	Flat	Long	Dia.	Face	12"	2
Siding, hardboard	Framing	2-21/2x0.115	Sc-7z	13/64"	Casing	Long	Ndl.	Face	—	—
Siding, hardboard battenboard	Framing	11/2x0.083	Sc-7z	9/64"	Casing	Long	Ndl.	Face	—	—

See Notes, Key to Nail Types and Abbreviations on later page.

ROOM ADDITIONS

Application		Nail Size Inches	Nail Type	Head		Point		Nail-ing	Spacing o.c.	Nails Per Joint
	Nailed Into			Diameter Type		Size	Type			
Fascia, 1"	Framing, rafter	2½x0.120	Sc-2g	9/32" Flat		Med.	Dia.	Face	—	2
Fascia, 2" lumber	Framing, rafter	3¼x0.135	Sc-2g	5/16" Flat		Med.	Dia.	Face	—	2
Roofing, built-up	Sheathing	3/4–1¼x0.105	Sq-30	15/16" Square		Med.	Dia.	Face	10"	—
Roofing, built-up	Poured gypsum	1½–1¾x0.120	Sq-31	15/16" Square		Med.	Dia.	Face	10"	—
Roofing, asphalt shingle	Sheathing	3/4–2 x0.120	St-10g	3/8" Flat		Blunt	Dia.	Face	—	2-3
Roofing, asphalt shingle	Sheathing	3/4–2x.120–.135	Dr-10	3/8" Flat		Blunt	Dia.	Face	—	2-3
Roofing, wood shingle	Sheathing	3/4–2x.105–.120	Dr-18	3/16" Flat		Blunt	Dia.	Face	—	2
Roofing, wood shingle	Sheathing	1¾–2 x0.083	St-18g	1/8" Flat, Csk.		Blunt	Dia.	Face	—	2
Roofing, asbestos shingle	Sheathing	As for siding								
Roofing, aluminum (corr. and flat)	Rafter, purlin	1½–1¾x0.145	Dr-10	13/32" Flat ★		Long	Dia.	Face	12"	—
Roofing, sheet metal (corr. and flat)	Rafter, purlin	1–3 x0.135	St-or	7/16" Flat ★		Long	Dia.	Face	12"	—
Roofing, glass fiber (corr. and flat)	Rafter, purlin	1½–3 x0.135	Sc-9g	7/16" Flat ★		Long	Dia.	Face	12"	—
Roofing, glass fiber (corr. and flat)	Rafter, purlin	1½–3 x0.148	Dr-9 or 10	★ With Neoprene washer attached		Long	Dia.	Face	12"	—
Lath, expanded metal, K-lath	Framing, joist	1½x0.148	St-23g	L-Shaped		Med.	Dia.	Face	6" & 12"	—
Lath, gypsum plasterboard	Framing, joist	1¼x0.101	St-23b	19/64" Flat, Csk.		Long	Dia.	Face	5"	—
Gypsumboard, 3/8"	Framing, joist	1¼x0.098	St-24	1/4"–19/64" Flat, Csk.		Long	Dia.	Face	5"–8"	—
Gypsumboard, 1/2"–5/8"	Framing, joist	1⅜x0.098	St-24	1/4"–19/64" Flat, Csk.		Long	Dia.	Face	5"–8"	—
Gypsumboard, prefinished	Framing, joist	1⅜x0.083	K-32e	3/16" Flat, Csk.		Long	Dia.	Face	5"–8"	—
Paneling, trim		1–1¼x0.054	K-32e	3/32" Casing		Blunt	Dia.	Face	—	—
Paneling, trim, exterior		1–1½x0.072	St-13	3/32" Casing		Blunt	Dia.	Face	—	—
Paneling, trim, exterior		1 x0.065	Sc-12	3/32" Casing		Blunt	Dia.	Face	—	—
Paneling, trim, exterior		1½x0.076	Sc-11	3/32" Oval		Blunt	Dia.	Face	—	—
Paneling, trim		1 x0.072	Sc-11	3/32" Casing		Blunt	Dia.	Face	—	—
Paneling, trim		1½–1¾x0.083	Sc-11	1/8" Casing		Blunt	Dia.	Face	—	—
Paneling, trim		2½x0.105	Sc-11	9/64" Casing		Blunt	Dia.	Face	—	—
Acoustic tile		1–1¾x0.062	St-25z	—		Blunt	Dia.	Face	—	—
Electric conduit	Wood	1½–2 x0.162	St-26z	1" Hook		Blunt	Ndl.	Face	—	—
Electric conduit	Masonry	1½–2 x0.162	St-27z	1" Hook		Blunt	Ndl.	Face	—	—
Fencing wire	Softwood (treated)	1½x0.148	St-22g	L-Shaped		Med.	Dia.	Face	—	—
Fencing wire	Hardwood	1½x0.148	St-21g	L-Shaped		Med.	Dia.	Face	—	—

See Notes, Key to Nail Types and Abbreviations on next page.

Key to Nail Types

Sc-iz	Screw-Tite Masonry Nail, hardened HCS, zinc plated
Sc-2	Screw-Tite Framing Nail, hardened HCS
Sc-2g	Screw-Tite Framing Nail, hardened HCS, galvanized
Sc-3	Screw-Tite Framing Nail, bright LCS
Sc-3g	Screw-Tite Framing Nail, bright LCS, galvanized
Sc-4	Screw-Tite Flooring Nail, hardened HCS
Sc-5	Screw-Tite Trussed Rafter Nail, hardened HCS
Sc-6g	Screw-Tite Asbestosboard Nail, hardened HCS, galvanized
Sc-7g	Screw-Tite Exterior Hardboard Nail, hardened HCS, galvanized
Sc-7z	Screw-Tite Exterior Hardboard Nail, hardened HCS, zinc plated
Sc-8	Screw-Tite Casing Nail, silver bronze
Sc-9g	Screw-Tite Roofing Nail, hardened HCS, galvanized
Sc-10g	Screw-Tite Roofing Nail, bright LCS, galvanized
Sc-11	Screw-Tite Finishing Nail, bright LCS,
St-12	Screw-Tite Finishing Nail, stainless steel
St-3	Stronghold Framing Nail, bright LCS
St-4	Stronghold Parquet Flooring Nail, hardened HCS
St-6g	Stronghold Asbestosboard Nail, hardened HCS, galvanized
St-9g	Stronghold Roofing Nail, hardened HCS, galvanized
St-10g	Stronghold Roofing Nail, bright LCS, galvanized
St-12	Stronghold Finishing Nail, stainless steel
St-13	Stronghold Finishing Nail, monel metal
St-14	Stronghold Sinker Nail, bright LCS
St-14g	Stronghold Sinker Nail, bright LCS, galvanized
St-15	Stronghold Underlay Nail, Hardened HCS
St-16	Stronghold Underlay Nail, bright LCS
St-17	Stronghold Sheathing Nail, bright LCS
St-18g	Stronghold Shingle Nail, bright LCS, galvanized
St-18ge	Stronghold Shingle Nail, bright LCS, galvanized and enameled
St-19t	Stronghold Shingle Nail, bronze, tin plated
St-20	Stronghold Shingle Nail, stainless steel
St-21g	Stronghold Fence Staple, hardened HCS, galvanized
St-22g	Stronghold Fence Staple, bright LCS, galvanized
St-22z	Stronghold Fence Staple, bright LCS, zinc plated
St-23b	Stronghold Lath Nail, bright LCS, blued
St-24	Stronghold Drywall Nail, bright LCS
St-25z	Stronghold Kollarnail, hardened HCS, zinc plated
St-26z	Stronghold Conduit Staple, bright LCS, zinc plated
St-27z	Stronghold Knurled Conduit Staple, hardened HCS, zinc plated
St-34	Stronghold Spike, hardened HCS
Hi-28	"Hi-Load" Shear-Resistant Nail bright LCS
Hi-17	"Hi-Load" Sheathing Nail, bright LCS
Sq-30	Squarehed Annular Thread Cap Nail, bright LCS
Sq-31	Squarehed Spiral Thread Can Nail bright LCS
K-32e	Annular Thread Kolorpin, bright LCS, enameled
Dr-8	Drive-Rite Spiral Thread Casing Nail, aluminum
Dr-9	Drive-Rite Screw Thread Roofing Nail aluminum
Dr-10	Drive-Rite Spiral Thread Roofing Nail aluminum
Dr-14	Drive-Rite Spiral Thread Sinker Nail, aluminum
Dr-20	Drive-Rite Spiral Thread Shingle Nail, aluminum
Dr-33	Drive-Rite Knurled Asbestos-Cement Shingle Face, aluminum

The above chart is based on a table appearing in Bulletin No. 38 (Revised Edition). "Better Utilization of Wood Through Assembly with Improved Fasteners," a study undertaken at Wood Research Laboratory, Virginia Polytechnic Institute, under the sponsorship of Independent Nail & Packing Company, Bridgewater, Mass., manufacturers of Stronghold® Annular Thread and Screw-Tite® Spiral Thread Nails and other improved fasteners.

STRONGHOLD ® ANNULAR THREAD NAIL

SCREW-TITE ® SPIRAL THREAD NAIL

STRONGHOLD ® SCREW THREAD NAIL

SCREW-TITE ® KNURLED MASONRY NAIL

NOTES: For fastening redwood, use only aluminum or stainless steel nails.

Local conditions, customs and popular usage may dictate minor variations in length and gauge of nails. Consult the Technical Service Department of Independent Nail & Packing Company, Bridgewater, Mass.

ABBREVIATIONS USED IN THIS TABLE:
Corr.–Corrugated
Dph. Diaphragm
Part. bd.–Particle board
LCS–Low Carbon Steel
HCS–High Carbon Steel
Plywd.–Plywood   Med.–Medium
Csk.–Countersunk   Dia.–Diamond
Flrg.–Flooring   Ndl.–Needle
Shgl.–Shingle

Copyright 1959, I. N. & P. Co. "Stronghold," "Screw-Tite," "Drive-Rite," "Squarehed," "Kolorpins," Trade Marks Reg. U.S. Patent Office.

ROOM ADDITIONS

Nails Required for Subflooring. The following table gives the recommended nailing schedule for helically-threaded nails. On wood flooring up to 4" wide, quantities are based on 8d flooring nails. For flooring 6" and wider, 10d nails have been figured. The quantities given below are sufficient to lay 1,000 b.f. of flooring:

Width Flooring	Joist Spacing 12" on centers	Joist Spacing 16" on centers
2"	40 lbs. 8d flooring	30 lbs. 8d flooring
3"	30 lbs. 8d flooring	23 lbs. 8d flooring
4"	22 lbs. 8d flooring	17 lbs. 8d flooring
6"	24 lbs. 10d common	18 lbs. 10d common
8"	17 lbs. 10d common	13 lbs. 10d common

Data on Common Wire Nails

Size of Nails	Length of Nails Inches	Gauge Number	Approximate Number per Lb.	Approximate Price per 100 lbs.
4d	1-1/2	12-1/2	316	$55.00
5d	1-3/4	12-1/2	271	55.00
6d	2	11-1/2	181	55.00
8d	2-1/2	10-1/4	106	51.00
10d	3	9	69	51.00
12d	3-1/4	9	63	51.00
16d	3-1/2	8	49	51.00

20d	4	6	31	51.00
30d	4-1/2	5	24	51.00
40d	5	4	18	51.00
50d	5-1/2	3	14	51.00
60d	6	2	11	51.00

ALUMINUM NAILS

Aluminum nails are excellent for use where nailheads are exposed to the atmosphere or corrosive conditions. Use aluminum nails whenever there is a possibility of rust or nail stain. Manufacturers usually do not recommend aluminum common nails for ordinary framing, because there is no economic advantage over steel wire nails for most common purposes, and the bending resistance of aluminum nails, despite the larger diameter, is less than that of steel nails.

Aluminum nails weigh about one-third as much as steel wire nails, but they are more expensive nail for nail. However, they do save labor and painting where rust or nail stain is a factor.

HARDWARE ACCESSORIES FOR WOOD FRAMING

The following items can be used for all types of wood construction. Developed primarily to provide better joints between wood framing members, in many cases the use of this hardware has resulted in lower overall costs—the additional material cost being more than offset by increased labor efficiency.

Steel Joist Hangers. Used for framing joists to beams and around openings for stairwells, chimneys, hearths and ducts, joist hangers are made of galvanized steel, varying from 12 ga. to 3/16" in thickness, with square supporting arms, holes punched for nails and with bearing surfaces proportioned to size of lumber. Approximate prices for sizes most commonly used are as follows:

ROOM ADDITIONS

Joist Size	Gauge	Depth of Seat	Opening in Hanger	Price Each
2" x 6"	12	2"	1-5/8" x 5"	$0.70
2" x 8"	12	2"	1-5/8" x 5"	0.78
2" x 10"	12	2"	1-5/8" x 8-1/2"	0.81
2" x 12"	10	2-1/2"	1-5/8" x 8-1/2"	1.00
4" x 8"	11	2"	3-5/8" x 5-1/4"	0.94
4" x 10"	9	2"	3-5/8" x 5-1/4"	1.12
4" x 12"	3/16	2-1/2"	3-5/8" x 8-1/2"	1.73

Framing Accessories. These are used in light wood construction to provide face-nailed connections for framing members. Adaptable to most framing connections, they eliminate the uncertainties and weaknesses of toe-nailing. They are made of zinc-coated, sheet steel in various gauges and styles. Framing accessories are designed to provide nailing on various surfaces. Special nails, approximately equal to 8d common nails but only 1-1/4" long, prevent complete penetration of standard nominal 2" lumber, are furnished with anchors. Approximate prices are as follows:

Type	Price per 100
Trip-L-Grips	$25.00
Du-Al-Clip	20.00
Truss Plates	30.00
Post Caps	66.00
H Clips	4.00
Angles	45.00

QUICK REFERENCE CHART #1
Labor Factors for Rough Framing

Item Description	Unit	Factor
Sills & Plates - Bolted	BF	0.02000
Studs	BF	0.02500
Floor Joists		
2" x 6"	BF	0.02200
2" x 8"	BF	0.02200
2" x 10"	BF	0.02000
2" x 12"	BF	0.02000
Girders - Built Up	BF	0.02000
Rafters - Gable		
2" x 4"	BF	0.03300
2" x 6"	BF	0.03000
2" x 8"	BF	0.03000
2" x 10"	BF	0.03000
2" x 12"	BF	0.03000
Rafters - Hip		
2" x 4"	BF	0.03600
2" x 6"	BF	0.03300
2" x 8"	BF	0.03000
2" x 10"	BF	0.03000
2" x 12"	BF	0.03300
Rafters - Flat		
2" x 6"	BF	0.02800
2" x 8"	BF	0.02600
2" x 10"	BF	0.02400
2" x 12"	BF	0.02400
Dormers	BF	0.04000
Beams/Girders - Heavy	BF	0.03000
Trusses	BF	0.06000
Bridging	BF	0.08000
Furring		
on masonry	LF	0.04000
on wood studs	LF	0.03000

ROOM ADDITIONS

Item Description	Unit	Factor	
Grounds			
on wood studs	LF	0.04000	
on masonry	LF	0.06000	
Wall Sheathing		**Diagonal**	**Horizontal**
1" x 6"	BF	0.02000	0.01600
1" x 8"	BF		0.01600
5/1 6" plywood	BF		0.01400
25/32" insulated	BF		0.01400
Subfloor		**Diagonal**	**Horizontal**
1" x 4"	BF	0.02000	0.01800
1" x 6"	BF	0.01700	0.01500
1" x 8"	BF	0.01600	0.01400
5/8" plywood	BF		0.12000
Roof Sheathing - Gable			
1" x 6"	BF	0.02000	
1" x 8"	BF	0.01800	
5/16" plywood	BF	0.01400	
open roof boards	BF	0.02000	
Roof Sheathing - Hip			
1" x 6"	BF	0.02200	
1" x 8"	BF	0.02000	
5/16" plywood	BF	0.01800	
Roof Sheathing - Flat			
1" x 6"	BF	0.01600	
1" x 8	BF	0.01400	
5/16" plywood	BF	0.01200	
1" x 8" horizontal	BF	0.02717	
1/2" plywood	BF	0.02046	
5/8" plywood	BF	0.02132	
3/4" plywood	BF	0.02217	
Columns			
8" x 8"	BF	0.06250	
10" x 10"	BF	0.04167	

Item Description	Unit	Factor
Scaffolding		
wood to 18'	BF	0.01316
	SF	0.01389
	LF	0.25000
steel tubular	SF	0.03000

Chapter 6

STAIRWAYS AND FLOORING

There are six basic categories of stairway: straight run, wide U, narrow U, wide L, double L and long L. In addition to these types, of course, there is an infinite variety of custom stairs.

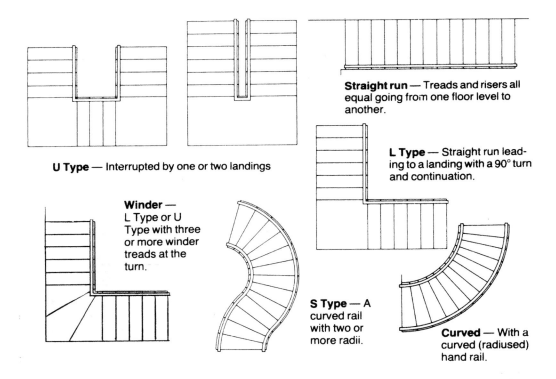

U Type — Interrupted by one or two landings

Straight run — Treads and risers all equal going from one floor level to another.

L Type — Straight run leading to a landing with a 90° turn and continuation.

Winder — L Type or U Type with three or more winder treads at the turn.

S Type — A curved rail with two or more radii.

Curved — With a curved (radiused) hand rail.

Today's homeowner demands that stairways be as pleasing to the eye as they are functional. Newels and balusters must fit the decor. Staircases must be wide enough to permit two people to pass comfortably and furniture to be carried up and down. The best width is 3 to 3-1/2 feet. They should also have a proper rise and run. If the rise is too steep, then it is tiring going up, dangerous coming down. If the run is too long, the foot may kick the riser at each step and the attempt to space steps is tiring and awkward.

Although there is a number of rules to determine rise and run, it is generally accepted that the riser be 7" to 7-1/2" high and the tread 10" to 11" wide. Another way to state it is that the sum of one riser and one tread should total between 17" and 18". For example, a riser of 7-1/2" plus a tread of 10" totals 17-1/2".

Headroom is also important. Even though the FHA minimum is 6'-8", studies indicate that it might not be sufficient. A more functional dimension if space is available is 7'-7", which provides more room for furniture and other household items to be moved upstairs and down with greater convenience. Service stairs are those that generally lead to a basement or an attic. They are usually steeper and made of less expensive materials.

Removing Existing Staircases. To remove an existing staircase, allow 3 to 5 hours, depending on the method of construction. If a railing is attached, it must be removed first, and then, any carpet must be lifted. If treads and risers are not dadoed into the stringers, one tread can be pried loose and the remainder removed from underneath. If the staircase is open and treads and risers are not dadoed, the entire stair can be removed after the ends of the stringers are loosened.

Where the stringers are part of an existing wall that is to be left intact, each stair must be cut in the middle so that the ends of the tread and riser can be pulled from the stringer. If the staircase was pre-built, stringers will most likely be damaged as the stairs are removed, and the stringers must be repaired before the stairs can be rebuilt. Allow 2 hours for the repair work.

	Hours	Rate	Total	Rate	Total
Carpenter	4.0	--	--	$25.60	$102.40

Framing and Installing Wood Stairs (Unassembled). To frame and erect ordinary wood stairs with closed stringers where the treads and risers are not housed, it will require 3 to 4 hours to lay out the job, cut out for stringers and risers and place stringers for a single flight of ordinary wood stairs up to 4'-0" wide and 11'-0" story height at the following labor cost:

	Hours	Rate	Total	Rate	Total
Carpenter	3.5	--	--	25.60	$89.60

Placing Treads and Risers. To place treads and risers, figure 2 to 3 hours at the following labor cost per flight, not including nosing, newels, balusters or handrail:

STAIRWAYS AND FLOORING

	Hours	Rate	Total	Rate	Total
Carpenter	2.5	--	--	$25.60	$64.00

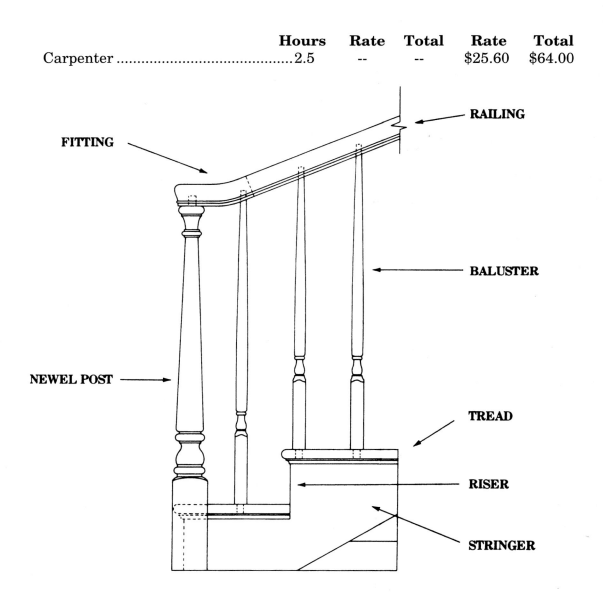

Placing Rails. To place 4" x 4" newels, 2" x 4" rails and 1" x 2" balusters for the above type of stairs, allow 2 to 3 hours time at the following cost per flight:

	Hours	Rate	Total	Rate	Total
Carpenter	2.5	--	--	25.60	$64.00

Erect Complete Staircase with Handrails Only. To lay out and erect a complete staircase, including plain wall handrail secured by brackets, figure 6 to 8 hours time at the following labor cost per flight:

	Hours	Rate	Total	Rate	Total
Carpenter	7.0	--	--	25.60	$179.20

To complete the above including newels, handrail and balusters, one side, figure 7 to 10 hours time at the following labor cost per flight:

	Hours	Rate	Total	Rate	Total
Carpenter	8.5	--	--	25.60	$217.60

SHOP-BUILT WOOD STAIRS

Placing Shop-built Stairs. To place shop-built stairs with stringers housed out to receive risers and treads, figure 6 to 8 hours stairbuilder time to lay out the work, set stringers, treads and risers (wall rail attached to brackets) at the following labor cost per flight:

	Hours	Rate	Total	Rate	Total
Stairbuilder	7.0	--	--	25.60	$179.20

Placing Short Flights. Stairs having two short flights with an intermediate landing will require 10 to 12 hours stairbuilder time to lay out work, place stringers, treads and risers and install plain handrail at the following labor cost:

	Hours	Rate	Total	Rate	Total
Stairbuilder	11.0	--	--	25.60	$281.60

Plain Box Stairs. To install a plain box stair (3' to 4' wide and 9' to 10' story height) using open stringers will require 6 to 8 hours stairbuilder time to lay out

STAIRWAYS AND FLOORING

work and place stringers, treads and risers (not including balustrade or handrail) at the following labor cost for a single flight:

	Hours	Rate	Total	Rate	Total
Stairbuilder	7.0	--	--	25.60	$179.20

Plain Box Stairs, Two Flights. Plain box stairs consisting of two short flights with an intermediate landing between stories will require 10 to 12 hours of stairbuider time to lay out the work and place stringers, treads and risers for each short flight of stairs at the following labor cost per flight:

	Hours	Rate	Total	Rate	Total
Stairbuilder	11.0	--	--	25.60	$281.60

Short Flights, Complete. The complete labor cost for two short flights of stairs, including newels, handrails and balusters per story height should be approximately as follows:

	Hours	Rate	Total	Rate	Total
Stairbuilder	16.0	--	--	25.60	$409.60

WOOD STAIR WITH OPEN STRINGERS

Open Stringer (Return Nosing). Wood stairs of the open stringer type that have treads with a return nosing projecting out beyond the face of the stringer will require about 12 to 14 hours of stairbuilder time to lay out the work and place stringers, treads and risers for one flight of stairs containing 16 to 18 risers at the following labor cost per flight:

	Hours	Rate	Total	Rate	Total
Stairbuilder	13.0	--	--	25.60	$332.80

If required to place newels, handrail and balusters, add approximately 8 to 9 hours stairbuilder time to the above at the following labor cost per flight:

	Hours	Rate	Total	Rate	Total
Stairbuilder	8.5	--	--	25.60	$217.60

Framing Intermediate Landings. A carpenter should be able to lay out, cut and place a 4' x 8' landing in about 4 to 6 hours at the following labor cost per sq. ft.:

	Hours	Rate	Total	Rate	Total
Cost per s.f.	0.16	--	--	25.60	$4.10

Placing Wood Handrail on Brackets. The recommended height for handrails is 34" on landings and 30" on stairs. A carpenter should set brackets and install about 130 to 140 lin. ft. of wall-hung wood rail per 8-hour day at the following labor cost per 100 lin. ft.:

	Hours	Rate	Total	Rate	Total
Carpenter	5.9	--	--	25.60	$151.04
Cost per l.f.			--		1.51

Placing Wood Handrail on Metal Balusters. A carpenter should place about 90 to 110 lin. ft. of wood handrail on metal balusters per 8-hr. day at the following labor cost per 100 lin. ft.:

	Hours	Rate	Total	Rate	Total
Carpenter	8.0	--	--	25.60	$204.80
Cost per l.f.			--		2.05

Labor Hours for Stairbuilding

Work Unit	Man-hours
Cut new well (not including plaster)	8.0
Exterior rear stairs, with rail (limit 16 risers)	16.0
Add for each additional riser	1.0
Add for each winder	1.0
Exterior front stairs, with rail (limit 7 risers)	8.0
Replace steps, per tread, with riser, no stringers	1.0
Landings 4'x7' (add for columns and piers)	4.0
Interior strips - carpenter made per run, plus well	20.0
Stair Repair	
Tread only	0.5
Winders	1.0

STAIRWAYS AND FLOORING

Work Unit	Man-hours
Replace carriage on existing stairs	
Straight runs, per riser	0.5
Winders, per riser	1.0
Minimum charge for carriages	4.0

Typical Inside Stair Material Prices

Starting Steps
   Quarter Circle .... Red Oak .... 4'-0" long $33.00
   Half Circle .... Red Oak .... 4'-6" long 42.00
   Bull Nose .... Red Oak .... 4'-6" long 38.00
   Scroll End .... Red Oak .... 4'-6" long 42.00

Stringer (not housed)
   3/4" x 11-1/4" x 8'-0" long .... Red Oak $28.00
                  x 10'-0" long .... Red Oak 33.00
                  x 12'-0" long .... Red Oak 50.00
                  x 14'-0" long .... Red Oak 56.00
                  x 16'-0" long .... Red Oak 64.00

Treads (not returned; for returns add $1.75)
   1-1/16" x 10-1/2" x 3'-0" long .... Red Oak $11.00
                    x 3'-6" long .... Red Oak 12.00
                    x 4'-0" long .... Red Oak 14.00
   1-1/16" x 11-1/2" x 3'-0" long .... Red Oak 14.50
                    x 3'-6" long .... Red Oak 16.00
                    x 4'-0" long .... Red Oak 19.00
                    x 5'-0" long .... Red Oak 23.00
                    x 6'-0" long .... Red Oak 29.00

Return Nosing
   1-1/8" x 1-3/4" x 1'-0" with 2" return .... Red Oak $5.00

Tread Nosing
   1-1/8" x 1-3/4" x 1'-2" long .... Red Oak 1.90
   1-1/16" x 1-1/8" x 1'-2" long .... Red Oak 1.75

Landing Tread Nosing
   1-1/16" x 3-1/2" x 3'-0" long .... Red Oak 2.50
                  x 3'-6" long .... Red Oak 2.75
                  x 4'-0" long .... Red Oak 3.10

Tread Brackets
   12" x 7-3/4" 3 ply ....Natural Birch......................2.00

Newel Posts   **Birch**   **Red Oak**
   3-1/4" x 3-1/4" x 3'-2" tapered, starting...$14.50 ....$13.50
                    x 3'-10" tapered, starting...16.50 ....16.50
                    x 5'-3" tapered, landing .....22.50 ....23.00
                    x 7'-0" tapered, angle ....26.50 ....27.00

Balusters   **Birch**   **Red Oak**
   15/16" round   30" L...............................$2.20 ....$2.50
   or square   33" ....2.30 ....2.75
                   36" ....2.40 ....2.85
                   39" or 42"..........................2.75 ....3.00

Railings--Natural birch, random length
   1-3/4" x 1-5/8" $1.70
   1-3/4" x 1-11/16" 1.95
   2-5/8" x 1-11/16" 2.10
   2-1/4" x 2-3/8" 4.25
   Rail Bolts and Plug..0.35

CUSTOM, SPIRAL, AND DISAPPEARING STAIRS

Installing Custom Stairs. If custom stairs are purchased preassembled, the installation is greatly simplified. After the opening is boxed in, or the upper level reinforced as necessary, the stairs can be set in place. Because of their size and weight, three people are usually required to complete the job. Once in place, all that is needed is to secure the stringers.

	Hours	Rate	Total	Rate	Total
Stairbuilder	2.0	--	--	25.60	$51.20
Two Helpers	2.0	--	--	19.45	38.90
					$90.10

Pre-cut custom stairs are also available and can be installed by a stairbuilder working alone. Stringers must be placed and secured, then treads and risers attached. Allow 6 to 8 hours to complete the installation.

STAIRWAYS AND FLOORING

	Hours	Rate	Total	Rate	Total
Stairbuilder	7.0	--	--	25.60	$179.20

The installation of pre-assembled or pre-cut custom stairs is complete when newels, balusters and rails are placed. A wall railing can also be used if necessary. Figure 2 to 3 hours for these items.

	Hours	Rate	Total	Rate	Total
Stairbuilder	2.5	--	--	25.60	$64.00

"Deluxe" All Wood Spiral Stairways

Red Oak, 13 Risers .... 4'-0" diameter $3966.00
                                      4'-6" diameter 4000.00

Carpet Grade Treads .... 5'-0" diameter $4170.00
                                    5'-6" diameter 4245.00
                                    6'-0" diameter 4445.00

Above prices include standard 1-3/4" square baluster with 1/4" chamfered edge, 12 carpet grade treads, center support post, platform for carpet wrap, circular rail, landing rail across one side of platform and all necessary hardware for installation. Hardware on spiral stairs comes exposed, with an option for hidden hardware at additional cost. Prices are for knocked down parts for self installation.

Landing Rail: $56.00 per foot. (Rail needed to enclose stair opening includes rail, balusters, trim and hardware.)

Additional:
  Newel Post: $56.00 each
  Turned Newel Post: Add $24.00 per foot
  Turned Balusters: Add $17.00 per baluster
  Additional Risers: Add $330.00 each
  Solid Oak Treads: Add $12.00 each
  Solid Oak Platform: Add $115.00
  Hidden Hardware: Add $128.75

"Standard" All Wood Spiral Stairways

Red Oak, 13 Risers .... 4'-0" diameter $3050.00
                                              4'-6" diameter 3075.00

Carpet Grade Treads .... 5'-0" diameter $3250.00
                                              5'-6" diameter 3330.00
                                              6'-0" diameter 3525.00

The above prices include standard 1-3/4" square baluster with 1/4" chamfered edge, 12 carpet grade treads, center support post, platform for carpet wrap, circular rail, landing rail across one side of platform and all necessary hardware for installation. Hardware on spiral stairs comes exposed, with an option for hidden hardware at additional cost. (Prices for additions are same as in "Deluxe" stairway.) Prices are for knocked down parts for self installation.

Landing Rail. $56.00 per foot, includes rail.
Additional:
Hemlock Tread Stock
   36" .... $9.15 ea
   42" .... 10.60 ea
   48" .... 14.00 ea
Hidden Hardware: Add $130.00

Spiral Metal Stairs. These stairs are finding increasing use in everything from cottages to luxury homes. Construction is standardized to meet most code and design requirements. To the basic steel structure can be added various types of treads, handrails and railings. The stairs are shipped knocked down, and assembly and installation is simple. No welding is required. Figure a stairbuilder can assemble and place one flight of spiral metal stairs in 12 hrs. at the following labor cost per flight:

	Hours	Rate	Total	Rate	Total
Stairbuilder	12.0	--	--	25.60	$307.20

Installing Disappearing Stairs. Disappearing stairs are primarily used for attics, where space for regular stairs is not available. Units come factory assembled ready to install. The standard sizes of disappearing stair openings are 22" x 54" and 25" x 54". Figure a carpenter can cut and frame the necessary opening in 2 to 3 hours at the following cost per opening:

STAIRWAYS AND FLOORING

	Hours	Rate	Total	Rate	Total
Carpenter	2.5	--	--	25.60	$64.00

The stair units are quite heavy, 100 to 150 lbs., and installation requires two carpenters. Allow 2 to 3 hours and two carpenters to install the unit properly at the following cost per installation:

	Hours	Rate	Total	Rate	Total
Carpenter	5.0	--	--	25.60	$128.00

Spiral Wood Stairs. Wood stairs come in 3'-6", 4'-0", 4'-6" and 5'-0" diameters, with heights made to fit the job. The balusters and center column are steel, but the treads, platforms and railings can be wood. Because the stairs are factory finished and assembled, two carpenters should be able to install the unit in one day at the following cost:

	Hours	Rate	Total	Rate	Total
Carpenter	16.0	--	--	25.60	$409.60

FLOORING

Hardwood Floors. Hardwood strip flooring is still one of the most beautiful, long-wearing floors a home can have. It comes in a wide variety of woods, the most common being oak, maple, beech, birch and pecan. The flooring is cut in narrow widths of varying thicknesses. The thinner strips are used primarily for remodeling when the floor is laid over an existing floor in order not to reduce room height. This is a benefit for the remodeler, because thinner strips cost less. He can offer the homeowner the beauty of wood at a lower price.

Oak is the most common hardwood flooring. Sizes vary from 1" to 3-1/2" in width, from 5/16" to 33/32" in thickness. In tongue and groove flooring the most commonly use is 25/32", which comes in 1-1/2", 2", 2-1/4" and 3-1/4", with 2-1/4" being the width most frequently installed.

For square-edge oak, the thickness is 5/16" in the following: 1", 1-1/8", 1-1/2" and 2" widths. Maple, beech, birch and pecan come in 25/32" and 33/32", with 2-1/2" and 3-1/2" widths in each thickness.

Grading - All Woods

Pecan. This flooring is manufactured in six standard grades, two specifying heartwood and one bright sapwood. Otherwise, color variation is not considered.

Maple, Beech, Birch. The grading of maple, beech and birch is almost identical. Neither sapwood nor varying colors is considered a defect in standard grades. These grades are first, second and third, with first being the best grade. Each of these grades comes in special color grades, selected for uniformity of color—*selected first grade light northern hard maple* and *selected first grade amber northern hard maple*.

Oak (Quarter Sawed). *Clear.* The face is practically clear, admitting an average of 3/8" of bright sap. The question of color is not be considered. Bundles are 2'-0" and up. Average length is 4-1/4'.

Select. The face may contain sap, small streaks, pinworm holes, burls, slight imperfections in working and small tight knots that do not average more than one for every 3 feet. Bundles to be 2'-0" and up. Average length is 3-3/4'.

Oak (Plain Sawed). *Clear.* The face should be practically clear, admitting an average of 3/8" of bright sap. Color is not be considered. Bundles are 2'-0" and up. Average length is 4-1/4'.

Select. The face may contain sap, small streaks, pinworm holes, burls, slight imperfections in working and small tight knots that do not average more than one to every 3 feet. Bundled, 2'-0" and up. Average length is 3-3/4'.

No. 1 Common. This grade will lay a good residential floor and may contain varying wood characteristics, such as flags, heavy streaks and checks, worm holes, knots and minor imperfections in working. Bundles are 2'-0" and up. Average length 2-1/2'.

No. 2 Common. This may contain sound natural variations and manufacturing imperfections. The purpose of this grade is to make available an economical flooring for homes, general utility use or where character and natural imperfections for appearance are desired. Bundles are 1-1/4' and up. Average length is 2-1/2'.

1-1/4 Shorts. Pieces are 9" to 18" long. Pieces graded No. 1 Common, Select and Clear are bundled together and designated as such. Pieces graded No. 2 Common are bundled and labeled separately. Although pieces 6" under or only 3" over the nominal length of the bundle may be included, pieces must averages 1-1/4' long.

STAIRWAYS AND FLOORING

Standard Thicknesses and Widths

Tongue and Groove		Square Edge	
Width	Thickness	Width	Thickness
25/32"	1-1/2", 2", 2-1/4", 3-1/4"	25/32"	7/8", 1", 1-1/8"
3/8"	1-1/2", 2"	5/6"	1-1/4", 1-1/3", 1-1/4", 2"

Removing Old Flooring

Removing Old Hardwood Strip Flooring. Figure one carpenter and one helper can remove 100 sq. ft. of flooring per hour at the following cost per sq. ft.:

	Hours	Rate	Total	Rate	Total
Carpenter	1.0	--	--	$25.60	$25.60
Helper	1.0	--	--	19.45	19.45
Cost per 100 s.f.			--		45.05
Cost per s.f.			--		0.45

To the above add 0.75 hours for removal of debris.

Removing Wood Subflooring. Where existing subfloors are badly worn and must be removed, figure one carpenter and one helper can remove 100 sq. ft. per hour at the following cost:

	Hours	Rate	Total	Rate	Total
Carpenter	1.0	--	--	$25.60	$25.60
Helper	1.0	--	--	19.45	19.45
Cost per 100 s.f.			--		45.05
Cost per sq. ft.			--		0.45

To the above add 0.5 hours for removal of debris.

Replacing Wood Subflooring. In small cut-up areas a carpenter should be able to lay between 700 and 800 b.f. of 1" x 6" or 1" x 8" at the following labor cost per 100 b.f.:

	Hours	Rate	Total	Rate	Total
Carpenter	1.00	--	--	$25.60	$25.60
Helper	0.75	--	--	19.45	14.59
Cost per 100 b.f.			--		40.19

Adding Plywood Over Existing Subfloors. Labor for 100 sq. ft. of plywood added over existing subfloor would be as follows:

	Hours	Rate	Total	Rate	Total
Carpenter	1.00	--	--	$25.60	$25.60
Helper	0.50	--	--	19.45	9.73
Cost per 100 s.f.			--		35.33
Cost per s.f.					--
					0.35

Estimating Material by Board Size

Flooring Size Inches	BF per 100 SF	1,000 BF will lay SF
25/32" x 1-1/2"	155.0	645.0
25/32" x 2"	142.5	701.8
25/32" x 2-1/4"	138.3	723.0
25/32" x 3-1/4"	129.0	775.2
3/8" x 1-1/2"	138.3	723.0
3/8" x 2"	130.0	769.0
1/2" x 1-1/2"	138.3	723.0
1/2" x 2"	130.0	769.0

Nails per 100 Sq. Ft.

Flooring	Nails	Spacing	Approx. Lbs. Per 100 Sq. Ft.
25/32" x 1-1/2" T&G	7d, 8d cut or screw or 2" barbed	10-12"	3.7
25/32" x 2-1/4" T&G	7d, 8d cut or screw or 2" barbed	10-12"	3.0
25/32" x 3-1/4" T&G	7d, 8d cut or screw or 2" barbed	10-12"	2.3

STAIRWAYS AND FLOORING

Flooring	Nails	Spacing	Approx. Lbs. Per 100 Sq. Ft.
1/2" x 2" T&G	5d cut or screw	8-10"	3.0
3/8" x 1-1/2" T&G	4d cut or screw	8"	3.7

Installing on Wood Subfloor

3/8" x 2" T&G	4d wire, cut or screwed or 1-1/4" barbed	6-8"	3.0
3/8" x 1-1/2" T&G	4d cut or screw	8"	3.7

Underlayment Prices

Plywood underlayment (Smooth one-side, "C" Cross Band)

Sq. Edge (5/8 P & TS)	0.65 per sf	20.80	4' x 8'
T&G (3/4)	0.80 per sf	25.60	4' x 8'

Particle board (underlayment)

4' x 8' - 3/8"	0.35 per sf
4' x 8' - 1/2"	0.36 per sf
4' x 8' - 5/8"	0.40 per sf
4' x 8' - 3/4"	0.45 per sf

Labor Laying 1000 B.F. Soft and Hardwood Flooring

	S.F. Laid 8-Hour Day	Installer Hours	Labor Hours
Ordinary Workmanship			
25/32" x 3-1/4" face softwood floors for porches, kitchens, etc.	500-600	14.5	4
25/32" x 2-1/4" face oak flooring	350-400	21.0	4
Same as above laid by experienced floorlayer	400-450	19.0	4
25/32" x 1-1/2" face oak flooring	250-280	30.0	5
Same as above laid by experienced floorlayer	350-400	21.0	5

	S.F. Laid 8-Hour Day	Installer Hours	Labor Hours
First Grade Workmanship			
25/32" x 2-1/4" face oak flooring	275-300	28.0	4
Same as above laid by experienced floorlayer	325-375	23.0	4
25/32" x 1-1/2" face oak flooring	175-225	39.0	5
Same as above laid by experienced floorlayer	275-300	28.0	5

Estimating Quantities of Wood Flooring*

Measured Size, inches	Inches	Bundle	% Waste	Multiply Area by	BF Flooring per 100 SF
1 x 2	3/8 x 1-1/2	24	33.33	1.33	133
1 x 2-1/2	3/8 x 2	24	25	1.25	125
1 x 2-1/4	25/32 x 1-1/2	12	50.33	1.50	150
1 x 2-3/4	25/32 x 2	12	37.5	1.375	137.5
1 x 3	25/32 x 2-1/4	12	33.33	1.33	133
1 x 4	25/32 x 3-1/4	8	25	1.25	125

*When estimating the quantity of wood flooring required for any job, take the actual number of sq. ft. in space to be floored and add allowances as shown above.

Laying Wood Floors. In large cities, workmanship is often categorized in three classes. The first is generally a carpenter who does nothing but lay flooring. In the second are floorlayers who also do nothing but lay floors, but because of their proficiency, they can generally lay more flooring, more efficiently than a carpenter/floorer. The last category refers to a floorlayer who is not only proficient in laying floors but offers a quality of workmanship required in high class buildings and residences. First grade workmanship means all flooring closely driven up, laid free of hammer marks, with all nails set.

Labor Laying 100 S.F. of Hardwood Floors

Description of Work	S.F. Laid 8-Hour Day	Installer Hours	Labor Hours
Ordinary Workmanship			
25/32" x 3-1/4" softwood for porches, kitchens, etc.	400-500	1.8	0.6
25/32" x 2-1/4" face oak	250-300	2.9	0.6

Same as above by experienced floorlayer........ 300-350 .... 2.5 .... 0.6
25/32" x 1-1/2" face oak.................................... 150-200 .... 4.5 .... 0.8
Same as above by experienced floorlayer........ 225-275 .... 3.2 .... 0.8
First Grade Workmanship
25/32" x 2-1/4" face oak.................................... 200-225 .... 3.8 .... 0.6
25/32" x 1-1/2" face oak flooring...................... 120-150 .... 6.0 .... 0.8

Laying Strip Flooring Over Concrete. Many homes today are built on concrete slabs because of the savings. The slab serves as the subflooring. A fast, economical way to provide nailing for strip wood flooring over concrete subfloors is to lay sleepers in mastic.

Clean the slab and prime the surface. Snap a chalk line at right angles to the direction the flooring will run and cover the lines with adhesive according to the manufacturer's recommendations. Bottom sleepers treated with a wood preservative are placed in the adhesive and nailed at 24" intervals. The bottom sleepers are covered with polyethylene film 0.004" thick, which is joined by lapping the edges over the sleepers. A second course of untreated 1" x 2" sleepers is laid over the bottom course of sleepers and nailed through the polyethylene film into the bottom sleeper, spacing nails 18" to 24" apart. Using appropriate flooring nails, the flooring is blind-nailed at right angles to the sleepers. A minimum of 1/2" to 1" clearance between flooring and wall should be left to allow for expansion.

Under ordinary circumstances an installer should prime concrete, spread mastic and place sleepers and film at a rate of 50 sq. ft. per hour. Labor for placing the strip wood flooring is the same as for wood subflooring.

Finishing Hardwood Floors

Most unfinished hardwood flooring comes smoothly surfaced from the factory, but scratches, nicks and blemishes will occur when it is laid. Sanding is usually required, and the work should always be let out to a specialist.

The floors are first sanded with a floor sanding machine. The edges are finished by a disc-type edging sander that is capable of sanding up to the base shoe or quarter round. For a high quality job, most manufacturers recommend at least four sandings, starting with No. 2 grit, going down to No. 1/2, No. 0 and finally, No. 00. However, the grits used will vary with the condition of the floor.

Sanding. *Ordinary Grade Workmanship.* The class of workmanship or number of operations governs production and cost, but where an ordinary grade of workmanship is required in average-sized rooms, an experienced operator with a good machine should finish and edge 800 to 900 sq. ft. per 8-hr. day at the following cost per 100 sq. ft.:

	Hours	Rate	Total	Rate	Total
Machine Expense	1.0	--	--	$5.00	$ 5.00
Sand Paper		--	--	1.00	1.00
Operator	1.0	--	--	25.60	25.60
Cost per 100 s.f.			--		31.60
Cost per s.f.			--		0.32

First Grade Workmanship. In residences where first grade workmanship is required to eliminate all irregularities and waves, nicks and scratches, four cuts are required: first, with No. 1-1/2 or No. 2 grit, then with No. 1/2 or No. 1 grit and the last two cuts with No. 0 grit. On work of this kind an experienced operator should finish and edge 400 to 500 sq. ft. of floor per 8-hr. day at the following cost per 100 sq. ft.:

	Hours	Rate	Total	Rate	Total
Machine Expense	1.75	--	--	$5.00	$8.25
Sand Paper		--	--	2.00	2.00
Operator	1.75	--	--	25.60	44.80
Cost per 100 s.f.			--		55.05
Cost per s.f.			--		0.55

Resurfacing Old Floors. *First Grade Workmanship.* There is little difference between resurfacing an old floor and surfacing a new floor, except that on the old floor one extra cut with a special open-faced abrasive is necessary to remove old varnish or paint. First grade workmanship generally requires four cuts. The amount of sq. ft. finished in one day will vary with the condition of the floor, anywhere from 300 to 1,000 sq. ft. per 8-hr. day. The following is an approximation of the cost per sq. ft. of finish flooring in old floors:

	Hours	Rate	Total	Rate	Total
Machine Expense	1.0	--	--	$5.00	$5.00
Sand Paper		--	--	2.00	2.00
Operator	1.0	--	--	25.60	25.60
Cost per hour			--		32.60
Cost per s.f. (300 s.f. per 8-hr. day)			--		0.88

STAIRWAYS AND FLOORING

Selecting the right kind of flooring is crucial to the success of a project. This wood floor with "pegs" is just the right touch for the vintage look of this room.

	Hours	Rate	Total	Rate	Total
(500 s.f. per 8-hr. day)......					0.52
(750 s.f. per 8-hr. day)......					0.35
(1,000 s.f. per 8-hr. day)...					0.26

STAINING AND SEALING

Staining. The first coat of stain or other finish should be applied the same day as the last sanding. Otherwise, the wood grain can rise and the subsequent finish will be slightly rough. If stain is used, it should be applied before the wood filler and any other finishes. A painter should be able to apply 1200 to 1600 sq. ft. of stain per 8-hr. day at the following cost per 100 sq. ft.:

	Hours	Rate	Total	Rate	Total
Stain	--	--			$2.50
Painter...0.7	--	--		24.50	17.15
Cost per 100 s.f.			--		19.65
Cost per s.f.					0.15

The cost of good stain runs from $8.00 to $15.00 per gallon, which covers about 800 sq. ft.

Wood Filler. Paste wood filler is customarily used to fill crevices and large pores usually found in hardwoods. For residential oak flooring, it is always recommended. It is usually applied after floor stains and seals, but always before other finished materials. *Important: It should be allowed to dry 24 hours before the next operation is begun.* Paste wood filler comes in 3-lb. cans with prices from $7.00 to $8.00. A 1-lb. can will normally cover about 100 sq. ft. A painter can usually apply about 1300 to 1400 sq. ft. of filler per 8-hour day at the following cost per 100 sq. ft.:

	Hours	Rate	Total	Rate	Total
Filler	--	--			$ 2.50
Painter...0.6	--	--		24.50	14.70
Cost per 100 sq. ft.			--		17.20
Cost per s.f.					0.17

Sealing. The homeowner wants a floor surface that is attractive, durable, easily maintained and can be retouched in worn areas without creating a patchy appearance. The three types of finish preferred are floor seal, varnish and shel-

STAIRWAYS AND FLOORING

lac. The advantage of floor seal is that it penetrates the fibers and cells of the wood, sealing the pores in such a way that they are highly resistant to dirt, water and stains. It wears only as the wood wears down. Floor seal comes in a variety of decorator colors, averaging $15.00 per gallon and with coverage ranging from 400 to 450 sq. ft. per gallon for the first coat, 650 to 750 sq. ft. for the second coat. An experienced painter can finish about 1,000 sq. ft. of natural finish floor per 8-hour day at the following cost per 100 sq. ft. (Color seals take more time, because the excess pigment must be wiped by hand.)

	Hours	Rate	Total	Rate	Total
Sealer (first coat)		--	--		$3.75
Painter	0.8	--	--	24.50	19.60
Cost per 100 s.f.			--		23.35
Cost per s.f.					0.23

Varnish averages $14.00 per gallon with coverage of 350 to 400 sq. ft. for the first coat and 450 to 500 sq. ft. for the second coat.

Prefinished Plank Flooring. Oak is the preferred wood, but plank floors can be obtained in other woods as well as various veneers. The most popular oak comes in 3/4" thickness, in widths of 3" to 8" with random lengths. Generally, the pieces are tongue and grooved, with square or matched ends. They are also produced with square edges or matched ends. Special distressed finishes and pegging are also available. Plain flooring in mixed widths will run about $4.00 per sq. ft.

Premium finishes will add as much as $2.00 per sq. ft. One carpenter should lay at least 150 sq. ft. per 8-hour day.

	Hours	Rate	Total	Rate	Total
Carpenter	8.0	--	--	$25.60	$204.80
Oak Planking		--	--		600.00
Nails		--	--		5.00
Cost per 150 s.f.			--		$809.80
Cost per s.f.			--		5.40

PARQUET FLOORING

Wood parquet is available in many designs and woods, including mahogany, oak, walnut, teak, cherry and maple. Patterns are available in *prime grade, character marked, squared and beveled* edges. Sizes run 9" x 9", 12" x 12", 16" x

16"…up to 36" x 36". Standard thickness is 5/16", with custom orders of 11/16" and 13/16". Prices range from $3.00 per sq. ft. for standard quality to $6.00 per sq. ft. for premium quality. Parquet flooring can be nailed over wood subfloors or old finish floors, or laid in mastic over concrete. It can also be laid in mastic over subfloors and existing wood floors. Where installation is over square edge board subflooring, one layer of 15-lb. asphalt saturated felt should be laid in standard linoleum paste to keep the mastic from seeping through joints. Expansion joints must be left around all perimeters of rooms according to manufacturer specifications. Cost per sq. ft. for 9" x 9" parquet flooring installed with mastic is:

	Hours	Rate	Total	Rate	Total
Oak Parquet, 100 s.f.		--	--	--	$350.00
3-1/2 lbs. nails		--	--		3.15
Floorlayer	5.0	--	--	$25.60	128.00
Helper	5.0	--	--	19.45	97.25
Cost per 100 s.f.			--		578.40
Cost per s.f.			--		5.78

For 8" x 8" parquet figure labor for 100 sq. ft. at approximately 6.5 hours for nailing and 3.5 hours for adhesive.

Laminated Oak Blocks. Laminated oak blocks are fabricated from 3-ply all-oak plywood and come factory finished in 9" x 9" x 1/2" squares. A 27 sq. ft. carton (48 pieces) will cost about $94.50. Installation is generally the same as for parquet flooring and the cost per sq. ft. installed would be the same as above taking into account the difference in cost between the two products.

TILE AND SHEET FLOORING

Resilient flooring includes asphalt tile, vinyl asbestos tile, rubber tile, cork and linoleum. Each product has specific performance characteristics and should be selected more for performance than for cost. Manufacturer recommendations should be carefully checked before installation.

For example, if it is to be installed on or below grade, linoleum, cork and certain vinyls should not be used. If grease is a problem, rubber, cork and asphalt tile should not be used. If resilience and quietness are desired, cork and rubber should be considered, but leave vinyl out. One of the advantages of sheet goods is that they do not shrink. On the other hand, some tiles will shrink and separate.

Asphalt Tile. Asphalt tile is usually furnished 12" x 12" in various colors and patterns. It is one of several types of tile that can be used on concrete and below

STAIRWAYS AND FLOORING

grade, such as basement floors, because it is not susceptible to the moisture and alkali always present in concrete in contact with the ground. Asphalt tiles are used where a low-cost flooring is desired. Price per 12" x 12" tile is about $0.20 to $0.65.

When applying resilient flooring over wood subfloors, the floor should be smooth and level and covered with a layer of lining felt bonded to the wood with an approved mastic (which averages $5.50 per gallon and covers approximately 133 sq. ft.). Lining felt (15-lb.) comes in rolls 36" wide by 50 yards long (or 432 sq. ft.) costing about $12.50 per roll or $0.82 per sq. yd. Figure one worker can lay 250 to 300 sq. ft. per hour.

In some cases resilient flooring requires an underlayment of plywood or particle board. The boards should be staggered at the joints and separated by about 1/32" to allow for expansion. Underlayment boards come in 4' x 8' x 3/8" and average $5.30 per board. One worker can lay 320 sq. ft. per 8-hour day.

When estimating the quantity of resilient flooring to be installed, always figure a percentage for waste. The allowable percentages are shown below:

Approximate Percentages for Waste

Up to 50 sq. ft.	14%
50 to 100 sq. ft.	10%
100 to 200 sq. ft.	7 to 9%
200 to 300 sq. ft.	6 to 7%
500 to 1,000 sq. ft.	4 to 5%
1,000 to 5,000 sq. ft.	4%
5,000 to 10,000 sq. ft.	3 to 4%
10,000 and up	1.5 to 3%

Primer and Asphalt Cement Required. An experienced tile setter should be able to work at the following quantities and rates:

Tile Size	Sq. Ft. per 8-Hour Day	Hours per 100 Sq. Ft.
9" x 9"	400-425	2.0
12" x 12" (1/2")	650-700	1.2

Square Foot Prices for Asphalt Tile*

12" x 12" .... $0.25-0.65 (+ 0.10 per tile for adhesive)

*Depends on color

Cost of Laying 100 Sq. Ft. Asphalt Tile
Over Concrete Floors

	Hours	Rate	Total	Rate	Total
0.6 Gal. Asphalt Primer		--	--		$ 9.00
105 Sq. Ft. Tile (12" x 12")		--	--		47.25
0.5 Gal. Asphalt Cement		--	--		9.00
Tile-setter	1.6	--	--	25.60	40.96
Cost per 100 s.f.			--		106.21
Cost per s.f.			--		1.06

If concrete primer is not required, deduct asphalt primer, asphalt cement, 0.4 hrs. labor and add 0.75 gal. of asphalt emulsion.

Cost of Laying 100 Sq. Ft. Asphalt Tile
Over Wood Subfloor

	Hours	Rate	Total	Rate	Total
0.75 Gal. Emulsion Adhesive		--	--		$ 6.00
12 Sq. Yds. Lining Felt		--	--		6.00
105 Sq. Ft. Tile		--	--		47.25
0.75 Gal. Asphalt Emulsion		--	--		13.50
Tile-setter	1.2	--	--	18.75	30.72
Cost per 100 s.f.			--		103.47
Cost per s.f.			--		1.03

VINYL TILE

Plastic. Vinyl plastic tile is a more expensive resilient flooring material. However, it is very stable, will not shrink or expand and is highly resistant to oils, fats, greases, acids, alkalis, detergents, soaps and most solvents. Subfloor requirements are the same as for other resilient floorings. It can also be laid over concrete floors, on grade or below grade using special adhesives. Sizes available include 9" x 9" and 12" x 12" tile.

Material costs vary according to pattern, ranging from $1.05 per sq. ft. for 1/16" embossed patterns to $8.00 per sq. ft. for 1/8" clear patterns. They come in 1/16", 3/32" and 1/8" thicknesses. Installation rates are about the same as those

STAIRWAYS AND FLOORING

for asphalt tile. However, you can figure a tile setter can lay 100 sq. ft. of 9" x 9" tile in approximately 2.4 hours. For 100 sq. ft. of tile figure 1.35 gals. of mastic.

Estimating Floor Tile Quantities

	Tile Required for			
Sq. Ft.	6" x 6"	9" x 9"	12" x 12"	9" x 18"
20	80	36	20	18
40	160	72	40	36
60	240	107	60	54
80	320	143	80	72
100	400	178	100	90
200	800	356	200	178

Vinyl Parquet and Plank. This elastomeric floor covering looks like wood and has a grain you can actually feel. It is for both new construction and remodeling. The interlocking parquet pieces come in two patterns and both the parquet and plank come in five different simulated woods of varying colors to fit a wide variety of decors. Both have a polymer coating giving them excellent resistance to abrasion, stains and cigarette burns.

The interlocking parquet modules come in two sizes, approximately 19-3/4" x 19-3/4" for the larger and 16-3/4" x 13-3/8" for the smaller. The approximate laying area for the former is 33 sq. ft. per case and for the latter 35 sq. ft. per case. All planks are approximately 60" long and come in three face widths: approximately 4.4", 6.4" and 8.4".

Installation can be made over most sound substrates that are smooth, clean and free from hydrostatic water or moisture problems. It is set in mastic with a special adhesive supplied by the manufacturer. A gallon of adhesive will cost approximately $30.00 and cover 150 to 200 sq. ft.

Approximate Prices for Elastomeric Floor Coverings

Tile	per Sq. Ft.
19-3/4" x 19-3/4" (unfinished)	$1.87
16-3/4" x 13-3/4" (unfinished)	1.87
Oak Strip Flooring	
3/4" x 2-1/4" (unfinished)	2.60
3/4" x 2-1/4" (finished)	3.90

Oak Plank
3/4" x 4.4" (unfinished) 7.25
3/4" x 6.4" (unfinished) 7.25
3/4" x 8.4" (unfinished) 7.25
Kentile, Briarwood
12" x 12" x 1/8" 7.45
4" x 36" x 1/8" 5.60
Kentile, Parquet
12" x 12" x 1/8" 5.45

An experienced floor layer should be able to install about 100 sq. ft. of parquet or plank per 8-hour day at the following cost per sq. ft.:

	Hours	Rate	Total	Rate	Total
Installer	8.0	--	--	25.60	$ 204.80
Tile		--	--		794.85
Adhesive	0.5 gal.	--	--		<u>5.50</u>
Cost per 100 s.f.			--		1,005.15
Cost per s.f.			--		10.05

SHEET FLOORING

Vinyl. Vinyl sheet flooring is one of the most popular floor coverings for modern residential use. There's a color and pattern to suit almost any decor. What's more, research and technology have all but eliminated the few drawbacks exhibited when vinyl was first introduced.

Today's products have exceptional stain and abrasion resistance. Installation can be at all grade levels. New backings resist alkali moisture and hydrostatic pressure. Some even inhibit the growth of mildew. New surface coatings up to 25-mils thick minimize wear and upkeep.

Installation time and costs have been reduced with the addition of 12' widths to complement the standard 6' width. And under certain conditions a number of vinyls can be loose-laid without adhesive.

When applying vinyl sheet, many installers install a new subfloor over the old surface before putting the new covering down. Not only does this eliminate the irregularities of old flooring, but the composition of some of the new subfloor materials help provide a better bond and surface for the new flooring. One of the

STAIRWAYS AND FLOORING

newer products for this purpose is Armstrong *Tempboard*®. It comes in two sizes, 3' x 4' and 4' x 4', both 1/4" thick.

Figure an experienced installer can place about 270 sq. ft. of *Tempboard* in 4 hours at the following cost per 100 sq. ft.:

	Hours	**Rate**	**Total**	**Rate**	**Total**
Installer	1.48	--	--	25.60	$37.89
Tempboard		--	--		240.00
Cost per 100 s.f.			--		277.88
Cost per s.f.			--		2.78

175-Gauge Vinyl Sheet Flooring. A recommended adhesive for vinyl sheeting is S235, which costs about $17.00 per gallon. Coverage is 150 to 175 sq. ft. per gallon. An experienced installer of sheet vinyl should place about 450 sq. ft. of flooring in 4 hours at the following cost per 100 sq. ft.:

	Hours	**Rate**	**Total**	**Rate**	**Total**
Installer	0.89	--	--	25.60	$22.78
Adhesive	3.0 gals.	--	--		51.00
Cost per 100 s.f.			--		73.78
Cost per s.f.			--		0.74

For loose-laying and stapling, figure about 150 sq. ft. per hr. at a cost of approximately $0.15 per sq. ft.

QUICK REFERENCE CHART #2
Labor Factors for Rough Framing

Item Description	Unit	Factor			
Balusters		30" SW	30" HW	42" SW	42" HW
Minumum	EA	0.2857	0.2857	0.2963	0.2963
Maximum	EA	0.3077	0.3077	0.3200	0.3200
Newels		Starting		Landing	
Minimum	EA	1.1429		1.6000	
Maximum	EA	1.3333		2.0000	
Railings		Minimum		Maximum	
Built-up HW	LF	0.1333		0.1455	
Subrail	LF	0.0727			
Risers		Softwood		Hardwood	
1" x 8"	LF	0.1212		0.1250	
Skirtboard					
1" x 10"	LF	0.1455			
1" x 12"	LF	0.1538			
Treads	Length	3'-0"	4'-0"	5'-0"	6'-0"
5/4 x 10 oak	EA	0.4444	0.4706		
5/4 x 12 oak	EA	0.4444			0.5714
2" maple	EA	0.4444		0.5000	
2-1/4" maple	EA	0.4444		0.5333	
Box Stairs	Height	2'-0"	4'-0"	6'-0"	8'-0"
Prefabricated					
3'-0" wide	flight	2.7200	3.4000	3.8857	4.5333
3'-6" wide	flight	3.2000	4.0000	4.5714	5.3333
4'-0" wide	flight	3.5200	4.4000	5.0285	5.8667
Prefab Rail					
w/balusters	riser	0.2666			
Basement Prefab - 3' wide					
w/open riser	flight	4.0000			

STAIRWAYS AND FLOORING

Item Description	Unit	Factor			
Open Stairs		2'-0"	4'-0"	6'-0"	8'-0"
Prefab RSC no rail					
3'-6" wide	flight	3.2000	4.0000	4.5714	5.3333
Prefab RSC w/rail					
3'-6" wide	flight	4.2667	5.1429	5.8014	6.6667
3/6 x 3/6 Landing	EA	4.0000			

Item Description	Unit	Factor	
Curved Stairs - Oak		9'-0"	10'-0"
Open one side w/rail			
3'-3" wide	flight	22.8571	22.8571
Open two sides			
3'-3" wide	flight	32.0000	32.0000

Item Description	Unit	Factor
Residential - Oak		
Prefab	flight	10.6667
Built-in-place	flight	36.3636
Spiral - Oak w/rail		
Prefab		
4'-6" diameter	flight	10.6667

Chapter 7

DRYWALL, PANELING AND CEILING TILE

Walls and Ceilings. When it comes to walls, some customers look for colors and aesthetic qualities, while others want to know the material options and the advantages of each. Color has become the big marketing tool. The trend is toward softer and lighter colors, which give a room a more spacious feeling and permit lighting with fewer fixtures. Energy conservation has led to less lighting in smaller living spaces, and wall coverings are a part of this change.

The message here is that remodeling contractors should keep abreast of decorator trends even though they tend to originate in new construction. Today's homeowners want their homes to reflect contemporary styling even though they may be remodeling only one room.

In the older home it isn't paneling or other wall treatments that the remodeling contractor has to deal with initially. It is plaster. Often they are asked what caused cracks in walls and ceilings and whether to repair, replace or re-face them. Many cracks are not due to workmanship, but faults in the material to which they are applied. These can be lumber shrinkage, foundation settlement, inadequate bracing and expansion and contraction.

However, because of poor workmanship, plaster can spall, crack and disintegrate. A common condition occurs when the white coat is applied over a base coat that has been allowed to dry out too much. When the white coat is applied, the base coat soaks the moisture from the finish coat and a spidery pattern of cracking occurs. If the surface of the plaster seems soft, the plaster has probably been oversanded. If cracks occur at random on the white coat, the coat was probably too thin.

Unless major defects are involved, structural cracks reach an equilibrium and can be successfully patched if the plaster on each side is not loose. If the crack is less than 1/2" wide, dig out an inverted "v" joint, moisten the adjacent areas to reduce suction and fill with prepared patching compound in two stages.

For larger cracks, the crack should be made large enough to accommodate a piece of metal lath. The resulting channel can then be filled with a three-coat application. For small patch jobs, figure a minimum of $75.00 for labor and materials.

Stripping Walls. To strip a masonry wall of plaster and leave ready to relath and replaster, figure one laborer will strip 60 to 70 sq. ft. per hour.

	Hours	Rate	Total	Rate	Total
Laborer	1.54	--	--	$21.50	$33.11
Cost per s.f.			--		0.33

Replastering Stripped Walls. To replaster a stripped masonry wall with 2-coat gypsum cement plaster, figure the cost per sq. yd. as follows:

	Hours	Rate	Total	Rate	Total
1100 lbs. gypsum plaster			--		$ 58.30
33 cu. ft. of sand			--		17.81
340 lbs. of hydrated lime			--		23.80
170 lbs. of gauging plaster			--		12.75
Plasterer	17.0	--	--	24.30	413.10
Laborer	10.0	--	--	20.10	201.00
Cost per 100 sq. yds.			--		$726.76
Cost per sq. yd.					7.27

Three coat work will cost about $10.71 per sq. yd. Work involving small areas with corner beads will run considerably more.

Relath and Replaster Stripped Walls. To add 3/8" gypsum lath and 2-coat gypsum cement plaster on stripped wood studs, figure the cost per sq. yd. as follows:

	Hours	Rate	Total	Rate	Total
900 sq. ft. of 3/8" gypsum lath			--		$118.90
7 lbs. nails			--		5.95
Lather	8.0	--	--	25.40	203.20
1000 lbs. gypsum plaster			--		53.00
25 cu. ft. of sand			--		15.00
510 lbs. fin. hydrated lime			--		34.60
Plasterer	16.0	--	--	24.30	388.80
Laborer	9.0	--	--	20.10	180.90
Cost per 100 sq. yds.			--		$1,000.35
Cost per sq. yd.			--		10.00

Beads and moldings should be figured separately. If wired or stapled to lath, figure 2 1/2 hours per 100 lin. ft.; if nailed to masonry, figure 3 hours per 100 lin.

ft. Where door and window casing beads are involved, figure one hour for each opening. Corner reinforcing at inside angles can be figured at about 500 lin. ft. per 8-hour day.

PLASTERING

Combined Labor and Material Costs*
Price per Yd. on Rooms 8' to 10' High
(no allowance for openings)

Charge one yard for each lineal foot of wall area	**per yard**
30 yards or under	$15.00
32 to 100 yards	14.00
over 100 yards	12.00
over stairways	20.00

Add for:

Patching Around Windows	**per window**
Single	$20.00
Mullion	30.00
Triple	40.00

Arches	**per arch**
3'	$100.00
4' to 5'	135.00
6' to 8'	150.00

Corner Bead	**per lin. ft.**
Outside Corners	0.40

Close Window Openings	**per single opng**
with Plaster	30.00

DRYWALL

Drywall generally refers to gypsum wallboard, but there are other types of wall finishes, such as plywood, board, hardboard and fiberboard. The benefit of all these types of material is that they require few tools for application, and when correctly applied, can cover any imperfections of the previous wall.

Required Panel Thicknesses for Various Framing

Framing	Plywood	Fiberboard	Paneling	Gypsum
16" o.c.	1/4"	1/2"	3/8"	3/8"
20" o.c.	3/8"	3/4"	1/2"	--
24" o.c.	3/8"	3/4"	5/16"	1/2"

Plywood Paneling. Prior to installation, panels should be stored in the room where they are to be applied, or stored in an area having a similar room temperature. They should be stored flat with scraps of wood separating each panel so that they can breathe and absorb room moisture, and they should be covered. The number of panels required for a room with an 8' ceiling can be determined by measuring the perimeter of the room and using the table shown below.

Estimating the Number of 4' x 8' Panels Required Based on Perimeter of Room

Perimeter	No. of Panels	Perimeter	No. of Panels
36'	9	60'	15
40'	10	64'	16
44'	11	68'	17
48'	12	72'	18
52'	13	76'	20
56'	14	92'	23

1. For example, if the room measures 14' x 20', the total perimeter is 68', which would require 17 panels.

2. For door and window openings, use the following deductions: doors deduct 1/3 panel; windows deduct 1/4 panel; and fireplaces deduct 1/2 panel.

3. Always use the next higher number of panels when the total perimeter is between the ranges shown in the table.

DRYWALL, PANELING AND CEILING TILE

Approximate Prices of Paneling

Georgia Pacific
4' x 8' – 1/8"

Atlantic Cherry	$12.88 ea.
Autumn Oak	5.96
Barnboard	9.94
Birch (unfinished)	24.35
Blanc Polare	11.87
Cedar Panel	8.19
Nut Brown Birch	8.74
Pecky Pecan	9.94
Rustic Honey Oak	22.86

Suspended Ceiling Panels

Decorator	$2.39 to 3.39 ea.
Vinyl Coated	3.44 ea.
Fire-rated Acoustical	2.77 to 5.66 ea.

Ceiling Grid Metal

Wall Molding (10')	2.29 ea.
Main Runners (12')	4.49 ea.
2' Cross Tees	0.84 ea.
4' Cross Tees	1.55 ea.
Hanger Wire (6')	0.29 ea.
Hanger Hooks	1.29 doz.

Ceiling Tiles (12" x 12")

Decorator Tiles	0.29 to 0.68 ea.
Vinyl Coated Tiles	0.45 to 0.50 ea.
Acoustical Tiles	0.44 to 0.51 ea.

Accessories

Furring Strips (1" x 3")	$0.08 lin. ft.
Tile Cement	5.29 qt.
	14.95 gal.
Staples (1000 per box)	2.15 box

WALL PREPARATION

Existing Walls. If the existing wall is in good condition, the paneling can be applied directly to it. In closed wall applications all studs are located and marked with a chalk line, which should be extended 2" to 3" onto the floor or ceiling. A level on a 2" x 4" should be used to get all markings plumb and straight. All ceiling and floor moldings, window and door moldings, light fixtures and wall plates should be removed. *Be sure electricity is turned off at the main control box.* Tape all exposed wire ends or use wire nuts.

Open Stud Walls. Be sure all studs are plumb and evenly spaced. This application generally requires a header at the top and bottom between studs and blocking in the center which helps prevent twisting and gives added wall support. Both should be nailed flush with the stud face. On perimeter walls, insulation batts and a vapor barrier should be installed before the paneling is applied. This is priced separately.

If the studs are not wide enough to form nailing pieces for the panels, a strip of wood, usually 1" x 1" is nailed to either side of the stud along its entire length and flush with the face of the stud. With studs spaced 16" on center, a carpenter should cut and place 225 to 275 lin.ft. per 8-hour day at the following cost per 100 lin.ft.:

	Hours	Rate	Total	Rate	Total
Carpenter	3.2	--	--	$25.60	$81.92
Cost per l.f.			--		0.82

Concrete and Plaster. On masonry and concrete walls, furring strips should be applied every 16" horizontally, allowing 1/4" between the ceiling and top strip and between the floor and bottom strip. Vertical strips are inserted every 48" to support the panel edges. To obtain a level surface, the furring must be shimmed with wood shims, which are held in place by nailing through the furring strips. With studs 16" on center, a carpenter should cut and place 300 to 350 lin. ft. of furring strips per 8-hour day at the following labor cost per 100 lin. ft.:

	Hours	Rate	Total	Rate	Total
Carpenter	2.5	--	--	$25.60	$64.00
Cost per l.f.			--		0.64

DRYWALL, PANELING AND CEILING TILE

Installing Panels. The shade and texture of panels vary. It is best to stand the panels in position against the wall before installation and rearrange them to achieve the most attractive sequence of color texture.

Before installation, each panel should be cut 1/4" shorter than actual ceiling height. Measure floor to ceiling height in several locations to detect any variations. If height varies no more than 1/4", take the shortest height, subtract 1/4" and cut all panels to that size. If height varies more than 1/4", each panel must be cut separately. Ceiling and floor moldings cover height variations of up to 2".

Place the first panel in the corner against the adjacent wall. Be sure it is perfectly plumb and the right edge is centered over the stud. If not, cut the other edge of the panel so it will. Often, the corner where you are to begin will be irregular. To allow for such irregularities, place the panel in a plumb position and scribe a line, using a compass, from top to bottom. Cut the scribe line with a coping saw. With the first panel in position, the remainder of the panels should also fall on stud centers.

When nailing to studs, use 4d finishing nails spaced 6" on panel edges and 12" in grooves. When nailing to furring, use 2d nails, 8" apart along edges and 16" apart elsewhere. Use matching color head nails or countersink and fill with matching color putty.

The actual cost of paneling will vary with the size and shape of the room, whether full size sheets are used, or considerable cutting and fitting is necessary. On straight work in large rooms, a carpenter should be able to fit and place approximately 100 sq. ft. or three 4' x 7' or 4' x 8' panels in about 1.6 hours at the following cost per sq. ft.:

	Hours	Rate	Total	Rate	Total
Carpenter	1.6	--	--	$25.60	$40.96
Cost per s.f.			--		0.41

Gypsum Drywall. This sheet material is gypsum that is faced on both sides with heavy grade protective paper. It is used for walls and ceilings in all types of buildings and takes all kinds of decorating including paint and wallpaper.

Standard Sizes and Thicknesses of Gypsum Wallboard

Thickness	Width	Length
1/4"	48"	6 to 12 ft.
3/8"	48"	6 to 12 ft.
1/2"	48"	6 to 14 ft.
5/8"	48"	6 to 14 ft.

Quarter-inch gypsum wall board is used as a lightweight, low-cost, utility wallboard. It is also used for curved surfaces or to cover existing wall surfaces. Three-eighths is used principally for repair or remodeling work or in double wall construction. Half-inch is used in new construction, and 5/8" Type "X" is used where a one-hour fire-rating is required or framing space is in excess of 1/2" wallboard limitation.

The edges along the length are tapered, and on some types, the ends are also tapered. This permits taping and filling the joint. Gypsum wallboard joints reinforced with tape and cement make a joint as strong as the wallboard itself.

Installing Gypsum Board Using Nails. Gypsum board can be installed horizontally or vertically with nails, screws or adhesives. It is best to apply the panels horizontally using full sheets, because it keeps the number of joints to a minimum. Joints should be made at windows and doors whenever possible. If it is not possible, the joints should be staggered. For 3/8" material, use a 4d (1-3/8" long) nail. If ring shank nails are used, a nail about 1/8" shorter will provide adequate holding power.

Nails should be spaced 6" to 8" with a minimum edge distance of 3/8". Nail spacing is the same for horizontal and vertical applications. Nails should always be driven so that they dimple the panel.

"Nail pops" are where the head begins to come out, causing a bubble, and occur for several reasons. Drying out of the framing is one, and this can be reduced if moisture content of the framing is less than 15% when the panel is applied. Another is movement of the panel itself. This can be minimized in one of two ways. The first is to use special screws. The second is by *double nailing*. The latter consists of nailing the panel to each stud 12" o.c. The top and bottom are fastened with one nail at each stud. Spacing around the perimeter is 7" o.c. The last step is to place a second nail 2" from each nail on the inner surface.

Installing Gypsum Board Using Screws. Drywall screws are the preferred method for fastening drywall these days. Hardly anyone uses nails unless they really have to. (Adhesives are sometimes used, but this is rare, and it only occurs where screws and nails are both impractical.) Using a screw gun or an electric drill with a magnetic screwdriver bit, this is a far more efficient way to hang drywall than nailing.

Type W screws are used to fasten drywall to wood studs. *Type S* is used for metal studs. If screws are used, the spacing is not more than 12" apart when the studs are 24 o.c. and 16" apart when the studs are 16" o.c.

DRYWALL, PANELING AND CEILING TILE

Laboring Placing Gypsum Wallboard. When placing 3/8" gypsum wallboard in average size rooms, two experienced drywallers should place about 2000 to 2200 sq. ft. of wallboard per 8-hour day at the following labor cost per 100 sq. ft.:

	Hours	Rate	Total	Rate	Total
Drywaller	0.76	--	--	$26.60	$20.22
Cost per s.f.			--		0.20

For a carpenter to do patch work using the same material, cut the preceding installation rate in half.

When placing 3/8" or 1/2" wallboard in small rooms or spaces requiring considerable fitting and cutting in proportion to the number of sq. ft. of board placed, two drywallers should be able to place 800 to 1000 sq. ft. per 8-hour day at the following cost per sq. ft.:

	Hours	Rate	Total	Rate	Total
Drywaller	1.78	--	--	$26.60	$47.35
Cost per s.f.			--		0.47

When fire-resistant 5/8" material is specified, two drywallers should place about 1600 sq. ft. per 8-hour day on straight wall and ceiling construction at the following labor cost per sq. ft.:

	Hours	Rate	Total	Rate	Total
Drywaller	1.0	--	--	$26.60	$26.60
Cost per s.f.			--		0.27

Drywall Material Prices

4' x 8' - 3/8"	$4.70 ea.
4' x 8' - 1/2"	4.75
4' x 10' - 1/2"	7.19
4' x 12' - 1/2"	8.47
Rock Lath (32 sq. ft. bundles, 16" x 48" pcs.)	3.99 bdl.
Moisture Resistant Drywall (4' x 8' - 1/2")	5.79 ea.
Fire Shield Drywall (4' x 8' - 5/8")	5.75

Drywall Accessories

Ready Mixed Joint Compound	5 gal. 7.25
Joint Compound (add water)	18 lb. 5.67
Joint Tape	60' roll 0.97
	250' roll 1.66
Drywall Corner Bead (8' lengths)	.... 1.15 ea.

Plaster Supplies

Gypsolite (80 lbs.)	$5.49
Finish Lime (50-lb.)	3.79
Slo-Set (100-lb.)	8.55
Wood Fiber (50-lb.)	3.19
Retarder (1 1/2 lb.)	1.09
Plaster Patch (5-lb.)	2.59
Mason's Sand (80-lb.)	2.29

Taping Joints. After the wallboard has been installed, joints and corners have to be finished. There is a number of types of tape on the market. Typically, it comes in perforated fiber strips 2-1/16" wide by 250 feet long. Compound is sold in various container sizes, usually 1- and 5-gallon buckets. For patchwork they come together as a set, in boxes containing 60 feet of tape and enough compound to cover 150 sq. ft. of wall.

It is applied in four steps. First the tapered channel is filled with compound. Tape is immediately pressed into the compound squeezing excess out from under the edges. When thoroughly dry, at least 24 hours, a second coat is applied over the tape and feathered several inches beyond. For best results a third coat is applied after the second has dried. This is also feathered out beyond the second coat so that the total width of the seam is 12" to 14". Joints should be sanded after they dry to ensure a smooth, inconspicuous joint. Nail heads are not taped. However, they should also be filled with compound and brought flush with the surface of the board. Two to three coats are desirable.

On an average job an experienced taper should apply and finish 200 to 225 lin. ft. per 8-hour day at the following labor cost per lin. ft.:

	Hours	Rate	Total	Rate	Total
Taper	8.0	--	--	$25.10	$200.80
Cost per l.f.			--		0.94

DRYWALL, PANELING AND CEILING TILE

Taping by Machine. Most professional drywall joint finishers use a taping machine for applying perforated tape to the joints. An average operator using a taping machine can apply tape up to five times faster than by hand, resulting in an overall 15% to 20% labor savings.

Some contractors set a minimum cost of approximately $50.00 for taping when patching during remodeling work. Average material costs in the Chicago area work out to $0.35 to 0.45 per lin. ft. for small jobs and $0.25 to 0.35 per lin. ft. for large jobs. This does not include labor.

CERAMIC TILE

Ceramic tile provides a durable, colorful surface that is virtually maintenance free. Its applications include interior and exterior finishes as well as countertops for functional and decorative purposes. Special uses include acid resistant and electrically conductive installations. Ceramic tile is available in many sizes and shapes.

Glazed Ceramic Wall Tile. This tile has an impervious surface fused onto the body of the tile. It comes in many colors.

Ceramic Mosaic Tile. Available either glazed or unglazed, this tile has a facial area of less than six square inches and is usually mounted on 1' x 2' sheets to facilitate setting.

Quarry Tile. This rugged tile is used primarily as a finish flooring (interior and exterior), where a long-wearing, easy-to-clean surface is desired.

Estimating Tile Quantities. It is necessary to stress the importance of relating tile costs to the specifications of each individual job. Glazed ceramic wall tile, for example, can be backmounted or unmounted and can be installed using conventional portland cement mortar, various types of adhesives or dry-set portland cement. All of these variations can affect cost, both material and labor, and the estimator should review job requirements carefully before making the estimate.

Ceramic tile is estimated by the square foot, while trim pieces such as base and cap are estimated by the lineal foot. The estimator should deduct door and window openings, but the trim pieces needed to finish such openings must be added.

The finished estimate should include costs for:
1. Delivery to the job site;
2. Accessory materials such as wire mesh, sand and cement for floor fill under ceramic tile;

3. Mixing and placing floor fill;
4. Direct labor cost of laying and cleaning the tile;
5. Overhead.

When ceramic tile is applied to a plaster base, a question arises. Who will do the plastering, the tile contractor or the plasterer? The estimator should check the building trade jurisdictional agreements in his area. If no information is available, assume that the tile contractor will plaster bathrooms, vestibules and small halls in private residences and that all other plasterwork will be done by a plastering contractor. The three commonly accepted methods for setting ceramic tile are:

Conventional Portland Cement Mortar Method. The tile is bonded with a layer of pure portland cement paste to a portland cement setting bed. This is done while the setting bed is still plastic. Wall tile must be soaked in water so that the water needed for curing is not absorbed from the paste. This is the traditional method, and it is the most costly.

Dry-Set Portland Cement Mortar Method. This method uses a dry curing portland cement mortar, produced through the use of water retaining additives. It has made ceramic tile installation cheaper and simpler. Dry-set is ideally suited for use with concrete masonry, brick poured concrete and portland cement plaster. It should not be used over wood or gypsum plaster. Labor costs are appreciably reduced when this method is used.

Water-Resistant Organic Adhesive. Organic adhesives can be used over smooth base materials, such as wallboard, plywood and metal. It should not be used, however, in wet areas over surfaces that are subject to moisture penetration. Labor productivity is comparable to that for dry-set mortar. For large jobs or standard jobs the labor to set a specific amount of tile can be estimated using the table shown below.

Labor Productivity
(sq. ft. per team day; team is one tile setter and one helper)

	Description	Face-Mounted	Back-Mounted	Unmounted
Conventional Mortar	Glazed Wall Tile	....	75-90 SF	55-65 SF
	Ceramic Mosaic Tile			
	walls	45-50 SF	55-65 SF	....
	floors	100-125 SF	100-125 SF	....
	Quarry Tile Floors	....	....	100-125 SF

DRYWALL, PANELING AND CEILING TILE

	Description	Face-Mounted	Back-Mounted	Unmounted
Dry-Set Mortar	Glazed Wall Tile	....	125-150 SF	100-125 SF
	Ceramic Mosaic Tile			
	walls	100-125 SF	120-140 SF	....
	floors	125-150 SF	125-150 SF	....
	Quarry Tile Floors	....	....	125-150 SF
Organic Adhesive	Glazed Wall Tile	....	150-175 SF	120-140 SF
	Ceramic Mosaic Tile			
	walls	120-140 SF	125-150 SF	....
	floors	150-175 SF	175-200 SF	....
	Quarry Tile Floors	....	....	....
	Cove or Base	65-75 lin. ft.		
	Cap	80-90 lin. ft.		

For smaller jobs such as those frequently encountered in remodeling, the labor can be estimated by using the *labor factor* from the table below and multiplying it times the labor cost per sq. ft. of a standard job.

For example, to determine the cost of setting 10 sq. ft. of 1" x 1" ceramic mosaic tile in a small room using adhesive and one tile setter, figure one worker can set approximately 100 sq. ft. per 8-hour day. Using a labor figure of $25.60 per hour, it will cost $204.80 per day to set 100 sq. ft. or $2.05 per sq. ft. To find the cost of the 10 sq. ft. multiply the $2.05 times the labor factor 1.75 which equals $3.59 per sq. ft., or $35.90 for 10 sq. ft.

Labor Factors for Small Quantities

		Labor Factor
Ceramic Mosaic Tile	Small Rooms	1.75
	Countertops	3.00
Glazed Ceramic Wall Tile	Small Rooms	1.50
	Mantel Fronts	2.00
	Mantel Fronts w/returns	1.90
Cove or Base	Small Rooms	1.25
Cap	Small Rooms	1.10

To estimate the cost of a small job, multiply the labor costs given in the applicable unit price development by the appropriate labor factor. Material cost is not affected.

Plastic Wall Tile. Plastic wall tile made of styron, polystyrene and other similar plastics is usually furnished in 4-1/4" x 4-1/4" and 9" x 9" tiles in many colors—plain, granite tone and marbleized. It can be used as wainscoting in bathrooms, kitchens, laundries and any place where other kinds of wall tile are used.

A smooth base must be provided such as plaster, gypsum wallboard or other materials with a smooth, hard surface. Plastic wall tile should never be applied over porous insulation board. It can be cut with a cutter similar to that used for shingles. To install 100 sq. ft. requires about 2-1/2 gallons of adhesive. After the base has been applied, an experienced worker should be able to apply about 100 sq. ft. of tile per 8-hour day at the following labor cost per sq. ft.:

	Hours	Rate	Total	Rate	Total
Tiler	8.0	--	--	$25.60	$204.80
Cost per sq. ft.			--		2.05

Metal Wall Tile. Wall tile of aluminum, with baked enamel finish in various colors, or tile of copper or stainless steel are used for the same purposes as plastic wall tile. They require the same type of base and are applied with adhesive. A worker will apply about 100 sq. ft. per 8-hour day, same as for plastic tile above.

CEILINGS

Ceiling tile and suspended ceilings offer excellent alternatives to gypsum drywall and plastering. Made from natural wood or cane fibers, ceiling tile has excellent acoustic and insulating properties and comes in a variety of sizes, styles and colors. They are factory finished and do not require painting or additional decoration. Upkeep is minimal for the homeowner.

Installation also poses no particular problems. The tile can be applied directly to old plaster, providing the surface is level and in good condition; to gypsum drywall; over lath; lath over open beams or suspended, using suspended ceiling grid metal. If it is cemented to gypsum wallboard, the wallboard should be at least 3/8" thick and nailed on 16" centers.

The most frequently used sizes are 12" x 12" and 12" x 24" in 1/2" thickness with interlocking tongue and groove joints. It also comes in two other sizes that are frequently used, 16" x 16" and 16" x 32" in 1/2" or 3/4" thickness.

DRYWALL, PANELING AND CEILING TILE

Removing Old Ceilings. To remove plaster from ceiling joists, figure one worker will clear about 50 sq. ft. per hour. If the ceiling is suspended and the hangers are also to be removed, figure about 30 sq. ft. per hour.

To strip a plastered ceiling of acoustical tile, figure one worker can clear 75 sq. ft. per hour. Figure about $21.50 per hour to clean up and remove debris.

To add new plaster to ceilings will run approximately the same as for walls, the extra labor in the ceilings being offset by the fact that there are few if any openings. For 3/8" gypsum lath and 2-coat gypsum plaster, it costs about $12.50 per sq. yd. in fairly large areas.

To Install New Ceiling Tile Over Plaster. The first step is to measure the length and width of the room to determine whether or not the tile size being used will fit without cutting. Generally, however, it will have to be cut. This can occur when either the length, width or both work out to an odd size. In order to get the best appearance, the border tiles on adjacent walls should be the same width.

To figure the size of the border tiles for either the length or width of a room, use the following formula (using 12" x 12" tiles). Assume the room is 15'-6" long. This leaves an odd dimension of 6 inches. Add 12 inches to the 6 inches making 18 inches and divide by 2, giving 9 inches. To get an even border on each side, you will need 14 full length tiles and two 9-inch tiles, one at each end. (16" x 16" tiles). Convert the length or width of the ceiling to inches and divide by 16. Repeat the previous procedure but add 16 inches to the extra inches instead of 12 and again divide by 2. Bear in mind that the end border should never be less than 6 inches.

When applying the first tile directly to an existing ceiling, it should be placed in the center of the room. However, the position of the first tile will vary, depending on the width of the border tiles. Follow the manufacturer instructions for positioning the first tile.

Adhesive is placed in each corner and the middle of the underside of the tile. When pressing the tile in place, it should be slid back and forth slightly to insure a good bond. Succeeding tiles should be slid together firmly to engage the tongue and groove, but without pressure. All border tile should engage the tongue and groove but without pressure and should be cut with a reverse bevel so that the exposed face fits flush against the wall.

To tile an existing plaster ceiling with 12" x 12" tile, figure two workers together can install 800 tiles per 8-hour day at the following cost per 100 sq. ft.:

	Hours	Rate	Total	Rate	Total
100 sq. ft. of tile...................................		--	--	$29.00	$ 29.00
1.75 gal. adhesive...............................		--	--	14.90	26.08
Installers 2.0		--	--	25.60	51.20
Cost per 100 s.f.				--	$106.28
Cost per s.f..				--	1.06

Installing Tile on Furring Strips. When installing tile on furring strips, the placement of the first two furring strips is important in order to provide a properly spaced nailing surface for the border tiles.

Using 1" x 3" or 1" x 4" furring strips, nail the first flush against the wall at right angles to the joists. The second should parallel the first, at a distance equal to the width of the border tile plus 1/2" for the nailing flange.

Taking the previous example, the border tile figured out to 9". Adding the 1/2" for the nailing flange, the position of the second furring strip would be 9-1/2" from the opposite wall. The last strip should be nailed flush to the wall.

For larger tile the same procedure is used, with the furring strips being attached at a distance equivalent to the dimensions of the tile. The labor to attach furring strips for tile is approximately the same as attaching furring strips for wall panel.

The tiles are installed by starting in the corner and working diagonally across the room. Each is fastened with 1-1/8" blue nails or 9/16" staples. Figure one worker can fasten 35 to 40 tiles per hour at the following labor cost per 100 sq. ft.:

	Hours	Rate	Total	Rate	Total
Installer...2.5		--	--	$25.60	$64.00
Cost per s.f..				--	0.64

Suspended Ceilings. Suspended ceilings are quickly and easily installed. One of the advantages is that they can be adjusted to fit any ceiling height. They also reduce sound from the floor above, provide insulating value and space for ceiling lights. There are different types of grid metal by which the ceiling is hung, but the labor rates to install are all about the same. The grid should fall 2" to 2-1/2" below the joists or ceiling in order to allow enough room to insert the panels.

The principal elements of the grid are the main runners, cross tees and wall angles. The main runners, usually 12' long, are placed at right angles to the joists. They are spaced 2' to 4' on center, depending on the panel size. Cross tees are also spaced 2' or 4' apart, depending on the panel size.

DRYWALL, PANELING AND CEILING TILE

The wire used to hang the runners and tees is spaced 4' along the main runner and on each side of the splices. A chalk line around the room perimeter at the ceiling height is used to determine the position of the wall angle.

Figure the amount of material to order by measuring the long and short walls and rounding the numbers up until the dimensions can be divided by two. Thus, a 12'-4" x 16'-6" is rounded up to 14' x 18'. Multiply the two dimensions to determine the number of sq. ft. of material required, in this case, 252 sq. ft. Figure one worker can hand 200 sq. ft. of grid per 8-hour day at the following cost per sq. ft.:

	Hours	Rate	Total	Rate	Total
Installer	8.0	--	--	$25.60	$204.80
Cost per s.f.			--		2.05

INTERIOR MOLDING

When estimating the labor cost of interior moldings, the grade of workmanship should be taken into consideration, because the amount of work performed will vary considerably. The following costs are broken down into *Ordinary Workmanship* and *first class workmanship*. Ordinary workmanship is the most common and is usually found in structures such as medium priced residences and cottages. First class workmanship is required in higher priced homes.

Removing Existing Wood Base. A carpenter and helper working together should remove a one-member wood base at the rate of 100 lin. ft. per hour:

	Hours	Rate	Total	Rate	Total
Carpenter	1.0	--	--	$25.60	$25.60
Labor	1.0	--	--	19.45	19.45
Cost per 100 l.f.	1.0	--	--		$45.05
Cost per l.f.			--		0.45

Placing Wood Base. *Ordinary Workmanship*. This cost varies with the size of the room and whether a single member or 2- or 3-member base is specified. Where there are 55 to 60 lin. ft. of 2-member base in each room without an unusually large amount of cutting and fitting, a carpenter should place 125 to 150 lin. ft. per 8-hour day at the following labor cost per 100 lin. ft. If there is an unusually large number of miters to make, such as required in closets and other small rooms, increase the costs:

	Hours	Rate	Total	Rate	Total
Carpenter	5.8	--	--	$25.60	$148.48
Labor	1.0	--	--	19.45	19.45
Cost per 100 l.f.	6.8	--	--		$167.93
Cost per l.f.			--		1.68

A carpenter should place almost as many lin. ft. of 3-member base, consisting of two base members and a carpet strip, as 2-member (consisting of one member and carpet strip). Where there are 50 to 60 lin. ft. of base in a room, a carpenter should place 110 to 130 lin. ft. per 8-hour day at the following labor cost per 100 lin. ft.:

	Hours	Rate	Total	Rate	Total
Carpenter	6.7	--	--	$25.60	$171.52
Labor	1.0	--	--	19.45	19.45
Cost per 100 l.f.	7.7	--	--		$190.97
Cost per l.f.			--		1.91

First Grade Workmanship. In average-size rooms, a carpenter should place 100 to 115 lin. ft. of 2-member hardwood base per 8-hour day at the following labor cost per 100 lin. ft.:

	Hours	Rate	Total	Rate	Total
Carpenter	7.4	--	--	$25.60	$189.44
Labor	1.0	--	--	19.45	19.45
Cost per 100 l.f.	8.4				208.89
Cost per l.f.					2.09

Where 3-member hardwood base is used in average-size rooms, a carpenter should place 85 to 100 lin. ft., which is about two ordinary rooms, per 8-hour day at the following labor cost per 100 lin. ft.:

	Hours	Rate	Total	Rate	Total
Carpenter	8.7	--	--	$25.60	$222.72
Labor	1.0	--	--	19.45	19.45
Cost per 100 l.f.	9.7	--	--		$242.17
Cost per l.f.			--		2.42

DRYWALL, PANELING AND CEILING TILE

On work of this class, the wood grounds should be straight, so that it will not be necessary to force the wood base to fit tight against the finished plaster wall. Where a single 1" x 4" pine base is to be fitted to straight runs, a carpenter should set around 200 lin. ft. per day.

	Hours	Rate	Total	Rate	Total
Carpenter	4.0	--	--	$25.60	$102.40
Cost per l.f.			--		1.02

Placing Wood Picture Molding. *Ordinary Workmanship.* Where just an ordinary grade of workmanship is required, a carpenter should place picture molding in 5 or 6 average rooms per 8-hour day. This is equivalent to 250 to 275 lin. ft. at the following labor cost per 100 lin. ft.:

	Hours	Rate	Total	Rate	Total
Carpenter	3.0	--	--	$25.60	$76.80
Labor	0.5	--	--	19.45	9.73
Cost per 100 lin. ft.	3.5		--		$86.53
Cost per lin. ft.			--		0.87

First Grade Workmanship. Where the wood picture molding must fit close to the plastered walls with perfect miters, a carpenter should place molding in 4 to 5 ordinary-sized rooms per 8-hour day. This is equivalent to 175 to 200 lin. ft. at the following labor cost per 100 lin. ft.:

	Hours	Rate	Total	Rate	Total
Carpenter	4.4	--	--	$25.60	$112.64
Labor	0.5	--	--	19.45	9.73
Cost per 100 l.f.	4.9	--	--		$122.37
Cost per l.f.			--		1.22

If the picture molding is placed in fireproof buildings having tile or brick partitions, it will be necessary to place wood grounds for nailing the picture molding, but in non-fireproof buildings, the nails may be driven into the plaster, because the nails will obtain a bearing in the wood studs or wall furring.

Placing Wood Chair or Dado Rail. *Ordinary Workmanship.* In large rooms or long, straight halls and corridors, a carpenter should fit and place 275 to 300 lin. ft. of wood chair rail per 8-hour day at the following labor cost per 100 lin. ft.:

	Hours	Rate	Total	Rate	Total
Carpenter	2.8	--	--	$25.60	$71.68
Labor	0.5	--	--	19.45	9.73
Cost per 100 l.f.	3.3	--	--		$81.41
Cost per l.f.			--		0.81

In small kitchens, pantries, closets, bathrooms and the like, a carpenter will place only 160 to 180 lin. ft. of chair rail per 8-hour day at the following labor cost per 100 lin. ft.:

	Hours	Rate	Total	Rate	Total
Carpenter	4.7	--	--	$25.60	$120.32
Labor	0.5	--	--	19.45	9.73
Cost per 100 l.f.	5.2	--	--		$130.05
Cost per l.f.			--		1.30

First Grade Workmanship. Where first grade workmanship is required, a carpenter should place 200 to 225 lin. ft. of chair rail per 8-hour day at the following labor cost per 100 lin. ft.:

	Hours	Rate	Total	Rate	Total
Carpenter	3.8	--	--	$25.60	$ 97.28
Labor	0.5	--	--	19.45	9.73
Cost per 100 lin. ft.	4.3	--	--		$107.01
Cost per lin. ft.			--		1.07

For first grade workmanship in small rooms such as kitchens, pantries and bathrooms, which require considerable cutting and fitting around cabinets, wardrobes and cases, a carpenter should place 120 to 135 lin. ft. of chair rail per 8-hour day at the following labor cost per 100 lin. ft.:

	Hours	Rate	Total	Rate	Total
Carpenter	6.3	--	--	$25.60	$161.28
Labor	0.5	--	--	19.45	9.73
Cost per 100 l.f.	6.8	--	--		$171.01
Cost per l.f.			--		1.71

Placing Wood Cornices. *Ordinary Workmanship.* Where 3- or 4-member wood cornices are placed in living rooms, reception rooms and dining rooms, a

DRYWALL, PANELING AND CEILING TILE

carpenter should place cornice in one average-size room per 8-hour day, which is equivalent to 50 to 60 lin. ft., and the labor cost per 100 lin. ft. would be as follows:

	Hours	Rate	Total	Rate	Total
Carpenter	14.5	--	--	$25.60	$371.20
Labor	2.0	--	--	19.45	38.90
Cost per 100 l.f.	16.5		--		$410.10
Cost per l.f.			--		4.10

First Grade Workmanship. Where it is necessary for the wood members to fit the plastered walls and ceilings closely, with all miters true and even, two carpenters working together should complete 1 to 1-1/4 rooms per day. This is at the rate of 35 to 40 lin. ft. per 8-hour day for one carpenter at the following labor cost per 100 lin. ft.:

	Hours	Rate	Total	Rate	Total
Carpenter	21.0	--	--	$25.60	$537.60
Labor	2.0	--	--	19.45	38.90
Cost per 100 l.f.	23.0		--		$576.50
Cost per l.f.			--		5.77

Placing Vertical Wood Panel Strips. When vertical wood panel strips, or *battens,* are nailed to plastered walls to produce a paneled effect, a carpenter should place 22 to 28 pcs. (175 to 225 lin. ft.) per 8-hour day at the following labor cost per 100 lin. ft.:

	Hours	Rate	Total	Rate	Total
Carpenter	4.0	--	--	$25.60	$102.40
Labor	0.5	--	--	19.45	9.73
Cost per 100 l.f.	4.5		--		$112.13
Cost per l.f.			--		1.12

Placing Wood Strip Paneling. *Ordinary Workmanship.* Where panels are formed of wood molding 1-1/2" to 2-1/2" wide, making it necessary to cut and miter both ends of each panel strip, the lin. ft. cost will vary with the size of the panels and the amount of cutting and fitting necessary. There is almost as much labor required on a 2'-0" x 3'-0" panel as on one that is 3'-0" x 6'-0", even though there are only half as many lin. ft. in the former.

On small panels up to 2'-0" x 4'-0", requiring 12 lin. ft. of molding, a carpenter should complete 9 to 11 panels, containing 110 to 135 lin. ft. of molding per 8-hour day at the following labor cost per 100 lin. ft.:

	Hours	Rate	Total	Rate	Total
Carpenter	6.5	--	--	$25.60	$166.40
Labor	0.5	--	--	19.45	9.73
Cost per 100 l.f.	7.0	--	--		$176.13
Cost per l.f.				--	1.76

On larger panels 3'-0" x 5'-0" to 4'-0" x 6'-0" in size, where each panel contains 16 to 20 lin. ft. of molding, a carpenter should complete 7 to 9 panels, containing 140 to 180 lin. ft. of molding per 8-hour day at the following labor cost per 100 lin. ft.:

	Hours	Rate	Total	Rate	Total
Carpenter	5.0	--	--	$25.60	$128.00
Labor	0.5	--	--	15.45	9.73
Cost per 100 l.f.	5.5	--	--		$137.73
Cost per l.f.				--	1.38

First Grade Workmanship. Where wood panel moldings are used over canvassed or burlap walls, with all strips plumb and level, fitting closely to the plaster walls with perfect miters, a carpenter should complete 7 to 9 small panels, requiring 90 to 115 lin. ft. of molding per 8-hour day at the following labor cost per lin. ft.:

	Hours	Rate	Total	Rate	Total
Carpenter	7.8	--	--	$25.60	$199.68
Labor	0.5	--	--	19.45	9.73
Cost per 100 l.f.	8.3	--	--		$209.41
Cost per l.f.				--	2.09

On larger panels 3'-0" x 5'-0" to 4'-0" x 6'-0", where each panel contains 16 to 20 lin. ft. of panel molding, a carpenter should complete about 6 to 8 panels, containing 120 to 150 lin. ft. of molding per 8-hour day at the following labor cost per 100 lin. ft.:

	Hours	Rate	Total	Rate	Total
Carpenter	6.0	--	--	$25.60	$153.60
Labor	0.5	--	--	19.45	9.73
Cost per 100 l.f.	6.5	--	--		$163.33
Cost per l.f.			--		1.63

Placing Wood Ceiling Beams. *Ordinary Workmanship.* In buildings where built-up ceiling beams are used, the labor costs will vary depending on the number of beam intersections in a room and the length of the beams. It is just as easy to erect a 12'-0" built-up beam as an 8'-0" one.

On average work a carpenter should place 35 to 45 lin. ft. of built-up wood beams per 8-hour day at the following labor cost per 100 lin. ft.:

	Hours	Rate	Total	Rate	Total
Carpenter	20.0	--	--	$25.60	$512.00
Labor	3.0	--	--	19.45	58.35
Cost per 100 l.f.	23.0	--	--		$570.35
Cost per l.f.			--		5.70

First Grade Workmanship. For wood ceiling beams in higher class buildings, a carpenter should place 30 to 35 lin. ft. per 8-hour day, at the following labor cost per 100 lin. ft.:

	Hours	Rate	Total	Rate	Total
Carpenter	25.0	--	--	$25.60	$640.00
Labor	3.0	--	--	19.45	58.35
Cost per 100 l.f.	28.0	--	--		698.35
Cost per l.f.			--		6.98

Typical Costs for Pine Moldings

The prices given below are for pine moldings, but the contractor show be aware that there are other types of millwork worth considering. One increasingly popular alternative, both for interior and exterior, is urethane millwork.

Molded urethane moldings can be made to look like wood, and the fact that it is installed very much the same as wood makes its use an easy transition for the contractor.

Divider (8')	$1.59 ea.
Cap (8')	1.59 ea.
Inside Corner (8')	1.59 ea.
Outside Corner (8')	1.59 ea.
Bathboard Molding (8')	1.89 ea.
Decorator Board Molding (8', vinyl wrapped)	2.95 ea.

White Pine (unfinished)

2-1/4" SL casing	$0.43 l.f.
2-1/4" oval casing	0.45 l.f.
2-1/4" OG casing	0.45 l.f.
2" comb. casing and base	0.20 l.f.
3" SL base	0.45 l.f.
3-1/4" OG base	0.49 l.f.
Comb. base and shoe	0.40 l.f.
1-3/8" SL stop	0.18 l.f.
1-1/4" OG stop	0.24 l.f.
3/4" cap mold	0.24 l.f.
Stool	0.64 l.f.
3/4" corner bead	0.21 l.f.
1-1/8" corner bead	0.44 l.f.
1-5/16" corner bead	0.55 l.f.
3/4" parting stop	0.15 l.f.
1-1/8" lattice	0.14 l.f.
1-3/8" lattice	0.17 l.f.
1-3/4" lattice	0.21 l.f.
Colonial Band	0.44 l.f.
1-5/8" econ. base	0.39 l.f.
3-1/4" crown	0.54 l.f.
3-5/8" crown	0.79 l.f.
3-1/4" cove	0.60 l.f.
2-1/4" cove	0.43 l.f.
1-1/2" cove	0.36 l.f.
3/4" cove	0.19 l.f.
3/8" quarter round	0.10 l.f.

DRYWALL, PANELING AND CEILING TILE

1/2" quarter round	0.14 l.f.
3/4" quarter round	0.22 l.f.
1-1/16" quarter round	0.41 l.f.
1/2" x 3/4" base shoe	0.16 l.f.
Flat screen mold	0.14 l.f.
Brick mold	0.47 l.f.
Drip cap	0.27 l.f.
1-1/16" full round	0.39 l.f.
1-5/16" full round	0.45 l.f.
2" modern rail	0.59 l.f.
2-1/4" chair rail	0.44 l.f.
1-3/8" panel mold	0.21 l.f.
1-1/4" bed mold	0.25 l.f.
1-5/8" bed mold	0.33 l.f.
3/4" x 3/4" baluster	0.18 l.f.
Steel Closet Rod	0.55 l.f.

Cellular

Corners (8')	$2.19 ea.
Cap (8')	2.19 ea.
Casing (7')	2.19 ea.
Cove (8')	2.19 ea.
Stop (8')	2.19 ea.
Base (8')	2.59 ea.

Colonial

Corners (8')	2.99 ea.
Cap (8')	2.99 ea.
Stop (8')	2.99 ea.
Base (8')	3.55 ea.
Casing (8')	3.55 ea.
Crown (8')	3.55 ea.

QUICK REFERENCE CHART #3
Labor Factors for Plastering

Item Description	Unit	Factor
Cornices		
Miter	LF	1.50000
Running	SF	0.16000
Ornaments		
2"	LF	0.08000
3"	LF	0.09500
4"	LF	0.12000
Beam		
Arrises	LF	0.06000
Soffits	SF	0.03500
Cement Base		
with cove	LF	0.11000
beveled top	LF	0.14000
without cove	LF	0.06000
beveled top	LF	0.09000
Marking for Tile Effect		
Squares	SF	0.01444
Broken Joint	SF	0.01667
Sand Finish		
Interior		
Floated	SF	0.00444
Troweled	SF	0.00667
Coffered Ceiling		
10 SF Coffer	SF	0.17500
15 SF Coffer	SF	0.16000
20 SF Coffer	SF	0.14500
Over	SF	0.13000

DRYWALL, PANELING AND CEILING TILE

Item Description	Unit	Factor
Acoustical Plaster		
1/4"		
2 coat	SF	0.03000
Vaulted Ceiling		
1 radius	SF	0.02000
Groined Ceiling		
2 radii		
Internal	SF	0.04000
External	SF	0.06667
Textured Finish		
Sand		
Fine	SF	0.01667
Coarse	SF	0.01500
Texture		
Light	SF	0.01333
Coarse	SF	0.01667
Putty		
Texture		
Light	SF	0.01333
Coarse	SF	0.01667
Portland Cement Mortar		
on cork		
3 coat	SF	0.05778
Arrises	LF	0.10000
Drywall		
Standard	SF	0.00361
Cut Up	SF	0.00691

QUICK REFERENCE CHART #4
Labor Factors for Interior and Exterior Trim

Item Description	Unit	Factor Softwood	Hardwood
Baseboard			
1-member	LF	0.06000	
2-member	LF	0.07000	
3-member	LF	0.08000	0.10000
Chair Rail	LF	0.05000	0.06000
Plate Rail			
2-member	LF		0.12000
3-member	LF		0.16000
Picture Mold	LF	0.05000	
Ceiling Mold	LF	0.06000	
7/8" Panel Strips	LF	0.08000	
Water Table	LF	0.04000	
Drip Cap	LF	0.04000	
Corner Boards	LF	0.05000	
Verge Boards	LF	0.04000	
Closed Cornice			
2-member	LF		0.08000
Boxed Cornice			
3-member	LF		0.12000
Wood Gutters	LF		0.10000

DRYWALL, PANELING AND CEILING TILE

QUICK REFERENCE CHART #5
Labor Factors for Interior Millwork

Item Description	Unit	Factor Softwood	Hardwood
Apron			
2"	LF	0.03200	
3-1/2"	LF	0.03640	
Astragal			
1-3/4"	LF	0.03140	
2-3/16"	LF	0.03330	
Band			
1-1/8"	LF	0.02960	
1-3/4"	LF	0.03200	
Baluster			
1-1/16"	LF	0.03330	
1-5/8"	LF	0.03640	
Base			
3-1/2"	LF	0.03330	
4-1/2"	LF	0.04000	
Bed			
1-3/4"	LF	0.02960	
2"	LF	0.03330	
Casing			
2-1/2"	LF	0.03330	
3-1/2"	LF	0.03720	
Chair Rail			
2-1/2"	LF	0.02960	
3-1/2"	LF	0.03330	
Closet Pole			
1-1/8"	LF	0.04000	
1-5/8"	LF	0.04000	
Cornice Molding			
1-3/4"	LF	0.02420	
2-1/4"	LF	0.02670	

Item Description	Unit	Factor					
Cornice Boards		2"	4"	6"	8"	10"	12"
1"	LF	0.0242	0.0320	0.0400	0.0400	0.0444	0.0444
3 member	LF	minimum 0.1000			maximum 0.1231		

Item Description	Unit	Factor
Cove		
1-3/4"	LF	0.02960
2-3/4"	LF	0.03140
Crown		
3-5/8"	LF	0.03200
4-5/8"	LF	0.03640
Casing-Door	SET	
1-1/8" plain		0.47060
1-1/8" detailed		0.47060

Item Description	Unit	Softwood	Hardwood
Cased Opening			
2-1/2" trim	EA	1.3559	
4-1/2" trim	EA	1.5094	
Glass Bead			
11/16"	LF	0.0281	
1/2"	LF	0.0291	
7/8"	LF	0.0296	
Half Round			
1/2"	LF	0.0296	
1"	LF	0.0314	
Handrail			
1-3/4"	LF	0.1000	
2-1/2"	LF	0.1053	
Lattice			
1-1/8"	LF	0.0296	
1-3/4"	LF	0.0320	
Milled Molding			
1"	LF	0.0296	0.0333
3"	LF	0.0333	0.0372
Parting Stop			
3/8" x 3/4"	LF	0.0291	
1/2" x 3/4"	LF	0.0314	

DRYWALL, PANELING AND CEILING TILE

Item Description	Unit	Factor Softwood	Hardwood
Quarter Round			
1/4"	LF	0.0291	
3/4"	LF	0.0314	
Stool			
11/16" x 3-1/4"	LF	0.0400	
1-1/8" x 3-1/4"	LF	0.0533	
Threshold			
Inside	EA		0.2500
Outside	EA		0.5000
Wainscot			
Minimum	LF	0.1053	
Maximum	LF	0.1231	
Full Bound w/stool			
Minimum	SET	0.6154	
Average	SET	0.8000	
Maximum	SET	1.3333	

Paneling	Unit	1/8"	1/4"
Temp Hardboard	SF	0.0320	0.0320
Temp Pegboard	SF	0.0320	0.0320
Moldings			
minimum	LF	0.0320	
maximum	LF	0.0376	
Plywood			
Birch			
minimum	SF		0.0320
average	SF		0.0381
maximum	SF		0.0457
Mahogany			
African	SF		0.0400
Philippine	SF		0.0320
Oak/Cherry			
minimum	SF		0.0320
maximum	SF		0.0400
Rosewood	SF		0.0500
Teak	SF		0.0400

Item Description	Unit	Factor	
		Softwood	Hardwood
Chestnut	SF		0.0427
Pecan	SF		0.0400
Walnut			
minimum	SF		0.0320
maximum	SF		0.0400

Chapter 8

PAINTING AND WALLPAPER

There is a variety of paints that can be used on interior walls. What kind one selects depends to a great extent on the room in which it is used, the surface to which it is applied and the wear and tear one can expect in a particular room.

Primers, undercoats and sealers not only improve the appearance of finish coats, but because they reduce the penetration of succeeding coats, they reduce the amount of paint needed. The best all-purpose wall paint is flat alkyd paint, because it will cover plaster, wood, wallboard, metal and wallpaper. It works well with one-coat application, goes on easy and is washable and odorless. Latex is another good paint, because it is easy to apply, dries quickly and brushes can be easily cleaned with water.

Some paints, even though they are labeled washable, may not be suitable for rooms where there is moisture, such as bathrooms, laundry rooms and even kitchens. Some flat no-sheen paints, whose color pigments lie near the surface, will hold moisture. Attempts to scrub away spots or stains can cause some of the pigment to come off leaving undesirable streaks.

There are different types of undercoat for use in various rooms and on various surfaces. On wood floors use a primer, followed by a floor seal and then floor or deck enamel. Wood paneling and trim should have a primer or undercoat and can be finished with latex, flat oil paint, semi-gloss paint, wood seal or varnish. Kitchen and bathroom walls also require a primer or undercoat and are best finished with semi-gloss paint. Drywall and plaster can be finished with latex or flat oil paint after they have been primed with a good primer or undercoat.

Dark and bright colors are harder to keep clean than pastels or whites, because scrub marks and scratches, water spots and streaks show up more on the former than they do on the latter. For medium or dark tone walls, choose a paint having a certain amount of gloss. High gloss is the most durable and moisture resistant. However, high-gloss paints tend to reflect glare spots of light, which many people find annoying. They might find that flat and semi-gloss paints are easier to live with.

Wood not only deserves but requires special consideration. Woods are divided into open-grain and closed-grain. For a smooth surface, open-grained woods require a paste filler before the wood can be finished. Birch, maple, gum and cherry require only a liquid filler. Close-grained woods don't require a paste filler, but they should be treated with a surface filler.

Close-Grained Woods	Open-Grained Woods
Fir	Ash
Gum	Beech
Pine	Butternut
Basswood	Chestnut
Redwood	Elm
Cedar	Mahogany
Cottonwood	Oak
Cyprus	Walnut
Poplar	
Apple	
Birch	
Boxwood	
Cherry	
Ebony	
Maple	
Pear	
Satinwood	

Paste fillers come ready-mixed from paint stores or building material suppliers and consist of silica, linseed oil, turpentine, driers and dyes. Tints can be added to achieve various colors. Liquid filler can also be purchased ready-mixed or made by thinning paste fillers with turpentine.

Paste fillers are applied with a stiff brush, brushing both ways, with the grain and across the grain. The excess is brushed away with a piece of coarse cloth or burlap, then allowed to dry for 24 hours.

The wood is then sealed with shellac or sealer made by thinning some of the top coat. A glazing coat is applied before the final top coat, if an antique finish is desired.

The top coat is brushed on in two coats, allowing appropriate drying time between coats. Sanding between coats with a 220 grit paper will insure a smooth surface.

ESTIMATING WOODWORK, WALLS AND CEILINGS TO BE PAINTED

In many cases, when figuring the amount of surface to be painted, you will be dealing with more than two dimensions. Certain surfaces and materials have to be factored in, so as to obtain the total dimensions of the surface. The following rules provide more precise dimensions.

PAINTING AND WALLPAPER

Picture Mold and Chair Rail. On picture mold and chair rail less than 6" wide, obtain the number of lin. ft. to be painted or varnished and figure 3/4 sq. ft. per lin. ft.

Wood and Metal Base. Wood or metal base, 6" to 1'-0" high, should be figured as 1'-0" high; under 6", it should be figured at 0.5 sq. ft. per lin. ft.

Wood Panel Strips and Cornices. When estimating quantities of wood panel strips, wainscot railing and ceiling cornices, measure the girth of the member. If the girth is less than 1'-0" wide, figure as 1'-0". If over 1'-0" wide, multiply the actual girth by the length: 1.25" x 200'-0" = 250 sq. ft.

Interior Doors, Jambs and Casing. When estimating quantities for interior doors, jambs and casing, add 2'-0" to the width and 1'-0" to the height of the opening. This allows for painting and varnishing the edges of the door, the door jambs, which are usually 6" wide, and the casing on each side of the door, which averages from 4" to 6" wide. Example: on a 3'-0" x 7'-0" door opening, and 2'-0" to the width and 1'0" to the height. This gives an opening of 5'-0" x 8'-0", containing 40 sq. ft. on each side. Some painters figure all single doors at 40 sq. ft. per side, or 80 sq. ft. for both sides, while others figure them at 50 sq. ft. per side or 100 sq. ft. for both sides. Do not deduct for glass in doors.

For a sash door that contains small lights of glass, add 2 sq. ft. for each additional light. A 4-lt. door would contain 8 sq. ft. additional; a 12-lt. door 24 sq. ft. additional; and so on.

Interior Windows, Jamb Linings, Sash and Casings. When estimating painting quantities for windows and window trim, add 2'-0" to the sides and length to allow for jamb linings, casing at the top and sides, and window stool and apron at the bottom. Example: If the window opening is 3'-0" x 6'-0", adding 2'-0" to both width and length gives a window 5'-0" x 8'-0" containing 40 sq. ft. of surface.

If the sash contain more than one light each, such as casement sash, add 2 sq. ft. for each additional sash. A 6-lt. window would contain 12 sq. ft. additional; a 12-lt. window, 24 sq. ft. additional.

Stairs. When estimating quantities of paint or varnish for wood stairs, add 2'-0" to the length of treads and risers to allow for stair stringers on each side of the stair. A stair tread is 10" to 12" wide, riser is about 7" high and the average cove and underside of a stair tread 2" in girth, so each tread should be figured 2'-0" wide. Multiply the width by the length to obtain the number of sq. ft. of paint or varnish required for each step. Multiply the area of each step by the number of treads in the stair, and the result will be the number of sq. ft. of finish. Example: A flight of stairs containing 20 treads, 4'-0" long. Adding 2'-0" to the length of the treads to allow for the stringers, gives 6 lin. ft. in each tread, which multiplied by the girth of 2'-0" gives 12 sq. ft. of surface in each step. There are 20 treads in the stairs, 12 x 20 = 240 sq. ft. For painting the soffits of stairs, use same rules of measurement as given above.

Balustrades and Handrails with Balusters. When estimating balustrades around well holes or stair handrail and balusters, measure the distance from the top of the treads to the top of the handrail and add 6" for painting or varnishing the handrail. Multiply the height of the balustrade by the length and the result will be the number of sq. ft. of balustrade. An easy method for computing the length of the stairs is to allow one lin. ft. in length for each tread in the stairs. For example, if the handrail is 2'-6" above the treads and the stairs contain 20 treads, add 6" to the height of 2'-6" to take care of the extra work on the handrail proper, making a total height of 3'-0". So, 3 x 20 = 160 sq. ft. of surface for one side of the balustrade or 320 sq. ft. if balustrade is on both sides.

The balustrade around the stairwell hole is easily estimated by multiplying the height by the length. On fancy balustrades having square or turned spindles, which require considerable additional labor, use actual measurements as given above and multiply by 4 to allow for extra labor.

Wood Ceilings. When estimating quantities for wood ceilings, multiply the length by the width. Do not deduct for openings less than 10'-0" x 10'-0".

Wainscoting. For plain wainscoting, obtain the actual area. For paneled wainscoting, obtain the actual area and multiply by 2.

Floors. To compute the quantities of floors to be finished, multiply the length of each room by the width, and the result will be the number of sq. ft. of floor.

Plastered Walls and Ceilings. To obtain the area of any ceiling, multiply the length by the breadth and the result will be the number of sq. ft. of ceiling. When estimating walls, measure the entire distance around the room and multiply by the room height. The result will be the number of sq. ft. of wall to be decorated. For instance, a 12'-0" x 15'-0" room has two sides 12'-0" long and two sides 15'-0" long, giving a total of 54 lin. ft. If the ceilings are 9'-0" high, 54 x 9 = 486 sq. ft. the area of the walls. Do not deduct for door and window openings.

Cases, Cupboards and Bookcases. When estimating surfaces of cupboards, wardrobes, bookcases and closets to be painted or varnished, obtain the area of the front and multiply by 3 if the cases do not have doors.

If the cases have doors, obtain the area of the front and multiply by 5. This takes care of painting or varnishing doors inside and out, shelves two sides and cabinet ends and backs.

Radiators. For each front foot, multiply the face area by 7.

PAINTING AND WALLPAPER

Sanding and Puttying Interior Trim. Sanding and puttying on high grade interior trim should be figured as one coat of paint or varnish; on medium grade trim, figure at 50% of one coat of paint or varnish.

COVERING CAPACITY OF PAINTS

Several factors, many of them extrinsic to the paint itself, influence covering capacity. It depends on how thin or thick the paint is brushed on. Dark paint, which hides the surface better than light paint, can be brushed out thinner. A rough surface requires considerably more paint than a smooth surface. Soft and porous wood absorbs more oil and requires more paint than close-grained lumber. One must consider the ingredients of the paint. Different brands of paint vary in covering capacity or hiding power.

Painting Interior Wood. New interior wood should receive three coats of paint—a priming coat, a second coat and a final or finish coat. Paint coverage varies with the type of work, such as doors and windows, running trim or paneling.

For doors and windows, coverage per gallon should be: first coat, 575 to 600 sq. ft.; second coat, 475 to 500 sq. ft.; third coat, 500 to 550 sq. ft.

For running trim up to 6" wide, coverage per gallon should be: first coat, 1100 to 1,200 lin. ft.; second and third coats, 800 to 900 lin. ft.

For ordinary flat wood paneling, coverage per gallon should be: first coat, 500 to 550 sq. ft.; second and third coats, 450 to 500 sq. ft.

When painting interior woodwork that has been previously painted and is in good condition, one gallon of paint should cover approximately the same as given for the second and third coats on new wood.

Painting Interior Walls. When painting interior walls, the covering capacity of the paint will depend on the surface—smooth or sand finished plaster, porous or hard finished wallboard. Three coats are recommended for interior plaster that has never been painted—a priming coat, a second or body coat and a third or finishing coat. However, if a two-coat job on unpainted plaster is desired, use a good wall primer or sealer for the first coat, followed by the second or finishing coat. To make two coats hide better, tint the first to nearly the same color as the second coat. If the surface has been painted before and the old paint is still in good condition, two coats are sufficient. No priming coat is required.

On smooth plaster or hard finished wallboard, one gallon of good wall primer or sealer should cover 575 to 625 sq. ft. for the priming coat. For the second coat, figure 500 to 550 sq. ft. per gallon and for the third coat, 575 to 625 sq. ft. per gallon.

On rough, porous sand finished plaster, one gallon of wall primer or sealer may cover only 275 to 300 sq. ft., while on very porous wallboard it may be only 150 to 200 sq. ft. per gallon.

For the second and third coats on rough surfaced plaster or wallboard, figure 400 to 475 sq. ft. per gallon.

One-Coat Wall Finishes. There are any number of one-coat wall finishes on the market, but when applied to new plastered surfaces or over surfaces that have been previously unpainted, it is advisable to use a primer or wall seal coat, which seals the pores in the plaster or other surface and provides a base suitable to receive the one-coat finish.

Latex Base Paints. There are several latex paints on the market that are compatible with most surfaces. When used over smooth plastered surfaces, one gallon should cover 450 to 500 sq. ft. for the first coat and 550 to 650 sq. ft. for the second coat. When applied over rough sand finish plaster, one gallon should cover 325 to 350 sq. ft. per coat.

Covering Capacity of Wall Size. The covering capacity of wall size varies greatly with the material used. Water size, which consists of ground or flake glue and water, covers 600 to 700 sq. ft. of surface per gallon. Varnish size consisting of varnish, benzine or turpentine and a little paint will cover 450 to 550 sq. ft. per gallon. Hard oil or gloss oil size, a mixture of rosin and benzine, covers 450 to 500 sq. ft. per gallon.

Covering Capacity of Wood Fillers. Liquid filler is normally used with close-grained woods such as pine and birch. One gallon should cover 500 to 550 sq. ft. It is necessary to use a paste filler to fill the pores of all open-grained wood such as oak, ash, walnut and mahogany. If paste filler is used on oak, one gallon of filler should cover about 450 sq. ft. of surface.

Covering Capacity of Shellac. The covering capacity of shellac varies with its purity. One gallon of good shellac should cover 550 to 750 sq. ft. of surface, depending on whether it is first, second or third coat.

Covering Capacity of Varnish. One gallon of good varnish should cover 400 to 450 sq. ft. of softwood floors with one coat; 200 to 225 sq. ft. with two coats and 135 to 150 sq. ft. with three coats.

When applied to hardwood floors, one gallon of good varnish should cover 500 to 550 sq. ft. with one coat; 250 to 275 sq. ft. with two coats; and 170 to 185 sq. ft. with three coats.

When applied to softwood interior finish, one gallon of good spar finishing varnish should cover 400 to 425 sq. ft. with one coat; 200 to 215 sq. ft. with two coats; and 135 to 145 sq. ft. with three coats.

PAINTING AND WALLPAPER

When applied to hardwood interior finish, one gallon of good spar varnish should cover 425 to 450 sq. ft. with one coat; 210 to 225 sq. ft. with two coats; and 140 to 150 sq. ft. with three coats.

Covering Capacity of Wax. One gallon of good liquid wax should cover 1,050 to 1,075 sq. ft. of surface.

Covering Capacity of Varnish Remover. The amount of varnish remover required to remove old varnish from floors and interior finish will depend on the condition of the old work, but one gallon of good varnish remover should be sufficient for 150 to 180 sq. ft. of surface.

Enamel Finish. Where interior woodwork is to receive an enamel finish, at least three coats are required as follows: oil base paint primer and sealer; second coat, prepared enamel undercoat; third coat, enamel finish coat. If a four-coat job is specified, the third coat may be a mixture of 1/2 undercoat and 1/2 enamel, followed by the enamel finish coat.

For flat work, figure material coverage per gallon as follows: paint primer and sealer, 575 to 600 sq. ft.; undercoat, 375 to 400 sq. ft.; enamel finish, 475 to 500 sq. ft.

For running trim, one gallon of material should cover as follows: paint primer and sealer, 1100 to 1200 lin. ft.; undercoat, 775 to 800 lin. ft.; enamel finish, 775 to 800 lin. ft.

Glazed Finish. One gallon of good glazing liquid, colored with oil colors, should cover 500 to 550 sq. ft. of surface.

For glazed finish on interior running trim, not over 6" wide, one gallon should cover 1050 to 1100 lin. ft.

ESTIMATING LABOR COSTS

In no other trade do labor costs vary so much as in painting and decorating. On some classes of work, the material costs are almost negligible, while the labor costs make it one of the most expensive kinds of work.

Cost of 100 Sq. Ft. (1 Sq.) 2-Coat Flat Finish Paint Applied to Smooth Plaster

	Hours	Rate	Total	Rate	Total
First Coat					
0.19 gal. paint	--	--	--	$14.50	$ 2.76
Painter	0.22	--	--	24.50	5.39

Second Coat
0.19 gal. paint .... -- -- 14.50 2.76
Painter..0.24 -- -- 24.50 5.88
Cost per 100 s.f. ....0.46 -- $16.79

Cost of 100 Sq. Ft. (1 Sq.) 2-Coat Flat Finish Paint Applied to Sand Finish Plaster

	Hours	Rate	Total	Rate	Total
First Coat					
0.21 gal. paint		--	--	$14.50	$ 3.05
Painter	0.27	--	--	24.50	6.62
Second Coat					
0.24 gal. paint		--	--	14.50	3.48
Painter	0.30	--	--	24.50	7.35
Cost per 100 s.f.	0.57		--		$20.50

Cost of 100 Sq. Ft. (1 Sq.) 3-Coat Industrial Enamel Applied to Smooth Plaster

	Hours	Rate	Total	Rate	Total
First Coat					
0.19 gal. primer		--	--	$17.50	$ 3.33
Painter	0.22	--	--	24.50	5.39
Second Coat					
0.24 gal. undercoat		--	--	14.50	3.48
Painter	0.25	--	--	24.50	6.13
Third Coat					
0.21 gal. enamel		--	--	18.80	3.95
Painter	0.28	--	--	24.50	6.86
Cost per 100 s.f.	0.75		--		$29.14

Cost of 100 Sq. Ft. (1 Sq.) Exterior Painting Applied from a Swing Scaffold

	Hours	Rate	Total	Rate	Total
0.17 gal. paint		--	--	$17.60	$ 2.99
Painter	1.6	--	--	24.50	39.20
Cost per 100 s.f.	1.6		--		$42.19

PAINTING AND WALLPAPER

Cost of 100 Sq. Ft. (1 Sq.) 2-Coat Oil Painted Applied to Smooth Metal Surfaces

	Hours	Rate	Total	Rate	Total
First Coat					
0.19 gal. paint	--	--	--	$18.80	$ 3.57
Painter	0.43	--	--	24.50	10.54
Second Coat					
0.18 gal. paint	--	--	--	18.80	3.38
Painter	0.38	--	--	24.50	9.31
Cost per 100 s.f.	0.81	--	--		$23.23

Cost of 100 Sq. Ft. (1 Sq.) 3-Coat Oil Paint Applied to New Steel Factory Sash

For old work, add 10% to 20% to prepare surfaces.

	Hours	Rate	Total	Rate	Total
First Coat					
0.10 gal. paint	--	--	--	$18.80	$ 1.88
Painter	1.18	--	--	24.50	28.91
Second Coat					
0.10 gal. paint	--	--	--	18.80	16.64
Painter	0.95	--	--	24.50	23.28
Third Coat					
0.08 gal. paint	--	--	--	18.80	1.50
Painter	0.77	--	--	24.50	18.87
Cost per 100 s.f.	2.90	--	--		$91.08

Paint costs vary according to quality used, color and local material prices. Add for surface preparation.

Cost of Sanding and Preparing 100 Sq. Ft. (1 Sq.) Interior Trim Before the First Coat

	Hours	Rate	Total	Rate	Total
Painter	0.35	--	--	24.50	$8.58

Cost of Sanding and Puttying 100 Sq. Ft. (1 Sq.) Interior Enamel Trim Between Coats

	Hours	Rate	Total	Rate	Total
Painter	0.80	--	--	24.50	$19.60

Cost of Light Sanding 100 Sq. Ft. (1 Sq.) Interior Enamel Trim Between Coats

	Hours	Rate	Total	Rate	Total
Painter	0.72	--	--	24.50	$17.64

Cost of 100 Sq. Ft. (1 Sq.) 4-Coat Enamel Applied to Interior Doors, Windows, Cases, Etc.

	Hours	Rate	Total	Rate	Total
First Coat					
0.17 gal. primer		--	--	$17.50	$ 2.98
Painter	0.70	--	--	24.50	17.15
Second Coat					
0.26 gal. undercoater		--	--	17.50	4.55
Painter	1.10	--	--	24.50	26.95
Third Coat					
0.11 gal. undercoater		--	--	17.50	1.93
Painter	1.10	--	--	24.50	26.95
Fourth Coat					
0.21 gal. enamel		--	--	18.80	3.95
Painter	<u>1.00</u>	--	--	24.50	<u>24.50</u>
Cost per 100 s.f.	3.90		--		$108.96

For preparatory work and sanding between coats, add as given above.
For three-coat work, omit cost of third coat.
For each additional coat, add same as fourth coat.

PAINTING AND WALLPAPER

Cost of 100 Lin. Ft. 4-Coat Enamel
Applied to Base, Chair Rail and Other Trim
Less Than 6 Inches Width

	Hours	Rate	Total	Rate	Total
First Coat					
0.09 gal. primer		--	--	$17.50	$ 1.58
Painter	0.40	--	--	24.50	9.80
Second Coat					
0.13 gal. undercoater		--	--	17.50	2.28
Painter	0.67	--	--	24.50	16.42
	1.07		--		$30.08

FINISHING INTERIOR WOODWORK

The greatest variations in cost will be found in interior finishing, because it is principally a labor proposition. For instance, where two, three or four coats of paint, enamel or varnish are specified, some painters will apply the entire number of coats without sanding between coats, regardless of specifications. This saves time but produces an inferior job.

To produce a first-class finish on interior woodwork, it is necessary to sand or rub down between coats to produce a smooth surface to receive the following coats. The finish coat will then be smooth and without rough spots, dust marks, etc. It is practically impossible to obtain this finish without sanding between coats.

The costs given on the following pages cover a customary, good residential job, but if an exceptionally high grade of workmanship is required, increase labor costs accordingly.

For all of the cost tables in this section, there are several rules that apply. For two coat work, deduct cost of third coat. For each additional coat of paint, add same as third coat. Do not forget to add for sanding and puttying as given below.

Cost of Sanding 100 Sq. Ft. (1 Sq.) Interior Trim
Before First Coat

	Hours	Rate	Total	Rate	Total
Painter	0.35	--	--	24.50	$8.58

Cost of Sanding and Puttying 100 Sq. Ft. (1 Sq.) Interior Trim Between Coats

	Hours	Rate	Total	Rate	Total
Painter	0.50	--	--	24.50	$12.25

Cost of 100 Sq. Ft. (1 Sq.) Light Sanding Interior Trim Between Coats

	Hours	Rate	Total	Rate	Total
Painter	0.30	--	--	24.50	$7.35

Cost of Backpriming 100 Lin. Ft. of Interior Trim Up to 6 Inches Wide

	Hours	Rate	Total	Rate	Total
0.09 gal. primer		--	--	$17.50	$1.58
Painter	0.21	--	--	24.50	5.15
Cost per 100 l.f.	0.21		--		$6.73

Cost of 100 Sq. Ft. (1 Sq.) 3-Coat Oil Paint Applied to Interior Doors, Windows, Cases, Etc.

	Hours	Rate	Total	Rate	Total
Priming Coat					
0.17 gal. paint		--	--	17.50	$2.98
Painter	0.70	--	--	24.50	17.15
Second Coat					
0.21 gal. paint		--	--	18.80	3.95
Painter	0.85	--	--	24.50	20.83
Third Coat					
0.19 gal. paint		--	--	18.80	3.57
Painter	0.85	--	--	24.50	20.83
Cost per 100 s.f.	2.40		--		$69.31

PAINTING AND WALLPAPER

Cost of 100 Lin. Ft. 3-Coat Oil Paint
Applied to Base, Chair Rail and Other Trim
Less Than 6 Inches Wide

	Hours	Rate	Total	Rate	Total
First Coat					
0.19 gal. paint		--	--	$17.50	$1.57
Painter	0.40	--	--	24.50	9.80
Second Coat					
0.12 gal. paint		--	--	18.80	2.26
Painter	0.60	--	--	24.50	14.70
Third Coat					
0.12 gal. paint		--	--	18.80	2.26
Painter	0.70	--	--	24.50	17.15
Cost per 100 l.f.	1.70	--			$47.74

Cost of 100 Sq. Ft. (1 Sq.) Spirit Stain
Applied to Interior Woodwork

	Hours	Rate	Total	Rate	Total
0.25 gal. of spirit stain		--	--	16.75	$ 4.19
Painter	0.50	--	--	24.50	12.25
Cost per 100 s.f.	0.50		--		$16.44

Cost of 100 Sq. Ft. (1 Sq.) Penetrating Oil Stain
Applied to Interior Woodwork

	Hours	Rate	Total	Rate	Total
0.15 gal. of oil stain		--	--	16.75	$ 2.51
Painter	0.50	--	--	24.50	12.25
Cost per 100 s.f.	0.50		--		$14.76

Cost of 100 Sq. Ft. (1 Sq.) Stain, Filler and Sealer, Wipe Off

	Hours	Rate	Total	Rate	Total
0.25 gal. of paste filler		--	--	$6.00	$ 1.50
Painter	1.50	--	--	24.50	36.75
Cost per 100 s.f.	1.50		--		$38.25

Cost of 100 Sq. Ft. (1 Sq.) Shellac
Applied to Interior Woodwork

	Hours	Rate	Total	Rate	Total
0.15 gal. pure shellac		--	--	12.00	$ 1.80
Painter	0.45	--	--	24.50	11.03
Cost per 100 s.f.	0.45		--		$12.83

Cost of 100 Sq. Ft. (1 Sq.) One Coat Gloss Varnish
Applied to Interior Woodwork

	Hours	Rate	Total	Rate	Total
0.23 gal. varnish		--	--	15.00	$ 3.45
Painter	0.60	--	--	24.50	14.70
Cost per 100 s.f.	0.60		--		$18.15

Cost of 100 Sq. Ft. (1 Sq.) Flat Varnish
Applied to Interior Woodwork

	Hours	Rate	Total	Rate	Total
0.16 gal. varnish		--	--	15.00	$ 2.40
Painter	0.60	--	--	24.50	14.70
Cost per 100 s.f.	0.60		--		$17.10

Cost of 100 Sq. Ft. (1 Sq.) Wax and Polish
Applied to Interior Woodwork

	Hours	Rate	Total	Rate	Total
0.18 gal. liquid wax		--	--	9.00	$1.62
Painter	0.95	--	--	24.50	23.28
Cost per 100 s.f.	0.95		--		$24.90

PAINTING AND WALLPAPER

Cost of Rubbing 100 Sq. Ft. (1 Sq.) Varnish Finish to Dull Rubbed Finish

	Hours	Rate	Total	Rate	Total
0.33 lb. powdered pumice stone		--	--	0.25	$0.08
0.13 gal. rubbing oil		--	--	1.50	0.20
1.00 lb. cotton waste		--	--	0.75	0.75
Painter	2.25	--	--	24.50	55.13
Cost per 100 s.f.	2.25		--		$56.16

Cost of 100 Sq. Ft. (1 Sq.) Interior Finish Consisting of 1-Coat Shellac and 1-Coat Varnish

	Hours	Rate	Total	Rate	Total
First Coat					
0.15 gal. pure shellac		--	--	12.00	$ 1.80
Painter	0.45	--	--	24.50	11.03
Second Coat					
0.23 gal. varnish		--	--	15.00	3.45
Painter	0.60	--	--	24.50	14.70
Cost per 100 s.f.	1.05		--		$30.98

Cost of 100 Sq. Ft. (1 Sq.) Interior Finish Consisting of 1-Coat Oil Stain, 1-Coat Shellac and 2-Coats Varnish

	Hours	Rate	Total	Rate	Total
First Coat					
0.15 gal. oil stain		--	--	16.75	$ 2.51
Painter	0.50	--	--	24.50	12.25
Second Coat					
0.15 gal. pure shellac		--	--	12.00	1.80
Painter	0.45	--	--	24.50	11.03
Third Coat					
0.23 gal. varnish		--	--	15.00	3.45
Painter	0.60	--	--	24.50	14.70
Fourth Coat					
0.23 gal. varnish		--	--	15.00	3.45
Painter	0.60	--	--	24.50	14.70
Cost per 100 s.f.	2.15		--		$63.89

Cost of 100 Sq. Ft. (1 Sq.) 2-Coat Synthetic Resin Finish Applied to Standing Trim

	Hours	Rate	Total	Rate	Total
First Coat					
0.17 gal. resin finish		--	--	$17.45	$ 2.97
Painter	0.50	--	--	17.52	12.25
Second Coat					
0.16 gal. resin finish		--	--	17.45	2.79
Painter	0.47	--	--	24.50	11.52
Cost per 100 s.f.	0.97		--		$29.53

FINISHING HARDWOOD FLOORS

Cost of 100 Sq. Ft. (1 Sq.) 2-Coat Paint Applied to Wood Floors

	Hours	Rate	Total	Rate	Total
First Coat					
0.19 gal. paint		--	--	$15.55	$ 2.95
Painter	0.35	--	--	24.50	8.58
Second Coat					
0.23 gal. paint		--	--	15.55	3.58
Painter	0.38	--	--	24.50	9.31
Cost per 100 s.f.	0.73		--		$24.42

Cost of 100 Sq. Ft. (1 Sq.) Paste Filler Applied to Hardwood Floors

	Hours	Rate	Total	Rate	Total
0.19 gal. paste filler		--	--	$6.00	$ 1.14
Painter	0.57	--	--	24.50	13.97
Cost per 100 s.f.	0.57		--		$15.11

Cost of 100 Sq. Ft. (1 Sq.) Varnish Applied to Hardwood Floors

	Hours	Rate	Total	Rate	Total
0.19 gal. varnish		--	--	$15.00	$ 2.85

PAINTING AND WALLPAPER

	Hours	Rate	Total	Rate	Total
Painter	0.33	--	--	24.50	8.09
Cost per 100 s.f.	0.33		--		$10.94

PAINTING AND DECORATING WALLS AND CEILINGS

Cost of Taping, Filling Joints and Sanding 100 Sq. Ft. (1 Sq.) Gypsum Wallboard

Based on approximately 1 lin. ft. of joint to 3 sq. ft. of sheetrock

	Hours	Rate	Total	Rate	Total
33.0 l.f. tape-cement		--	--	$ 0.04	$ 1.32
Painter	1.00	--	--	24.50	24.50
Cost per 100 s.f.	1.00		--		$25.82

Cost of 100 Sq. Ft. (1 Sq.) Sizing Applied to New Smooth Finish Walls

	Hours	Rate	Total	Rate	Total
0.16 gal. size		--	--	$ 6.00	$0.96
Painter	0.27	--	--	24.50	6.62
Cost per 100 s.f.	0.27		--		$7.58

Cost of 100 Sq. Ft. (1 Sq.) Prepared Wall Primer or Sealer Applied to Smooth Plaster or Dense Wallboard

	Hours	Rate	Total	Rate	Total
0.17 gal. sealer		--	--	$5.50	$ 0.94
Painter	0.45	--	--	24.50	11.03
Cost per 100 s.f.	0.45		--		$11.97

If porous insulating fiber wallboard, figure 0.5 gal. of primer per 100 sq. ft.

Cost of 100 Sq. Ft. (1 Sq.) Prepared Wall Primer or Sealer To Sand Finish Plaster

	Hours	Rate	Total	Rate	Total
0.35 gal. sealer		--	--	$ 5.50	$1.93
Painter	0.80	--	--	24.50	19.60
Cost per 100 s.f.	0.80		--		$21.53

Cost of 100 Sq. Ft. (1 Sq.) 1-Coat Sealer, 2-Coat Flat Wall Paint, Stippled, Applied to Smooth Plaster

	Hours	Rate	Total	Rate	Total
First Coat					
0.17 gal. sealer		--	--	$5.50	$ 0.94
Painter	0.45	--	--	24.50	11.03
Second Coat					
0.19 gal. paint		--	--	14.50	2.76
Painter	0.60	--	--	24.50	14.70
Third Coat					
0.17 gal. paint		--	--	14.50	2.47
Painter	0.55	--	--	24.50	13.48
Stippling					
Painter	0.52	--	--	24.50	12.74
Cost per 100 s.f.	2.12		--		$ 58.12

Cost of 100 Sq. Ft. (1 Sq.) 1-Coat Sealer, 2-Coat Flat Wall Paint Applied to Sand Finish Plaster

	Hours	Rate	Total	Rate	Total
First Coat					
0.35 gal. sealer		--	--	$5.50	$ 1.93
Painter	0.80	--	--	24.50	19.60
Second Coat					
0.24 gal. paint		--	--	14.50	3.48
Painter	0.70	--	--	24.50	17.15
Third Coat					
0.22 gal. paint		--	--	14.50	3.19
Painter	0.80	--	--	24.50	19.60
Cost per 100 s.f.	2.30		--		$64.95

Cost of 100 Sq. Ft. (1 Sq.) 1-Coat Sealer, 2-Coat Gloss or Semi-Gloss Oil Paint Applied to Smooth Plaster

	Hours	Rate	Total	Rate	Total
First Coat					
0.17 gal. sealer		--	--	$5.50	$ 0.94
Painter	0.45	--	--	24.50	11.03

PAINTING AND WALLPAPER

Second Coat

0.19 gal. paint		--	--	18.80	3.57
Painter	0.57	--	--	24.50	13.97

Third Coat

0.19 gal. paint		--	--	18.80	3.57
Painter	0.70	--	--	24.50	17.15

Stippling

Painter	1.10	--	--	24.50	<u>26.95</u>
Cost per 100 s.f.	2.82		--		$77.18

Cost of 100 Sq. Ft. (1 Sq.) 1-Coat Sealer, 2-Coat Gloss or Semi-Gloss Oil Paint Applied to Sand Finish Plaster

	Hours	Rate	Total	Rate	Total
First Coat					
0.35 gal. sealer		--	--	$5.50	$ 1.93
Painter	0.80	--	--	24.50	20.34
Second Coat					
0.26 gal. paint		--	--	18.80	4.89
Painter	0.83	--	--	24.50	20.34
Third Coat					
0.24 gal. paint		--	--	18.80	4.51
Painter	0.80	--	--	24.50	<u>19.60</u>
Cost per 100 s.f.	2.43		--		$70.87

Cost of 100 Sq. Ft. (1 Sq.) Glazing and Mottling Over Smooth Plaster

0.10 gal. glazing liquid		--	--	$ 6.50	$ 0.65
Painter	1.15	--	--	24.50	<u>28.18</u>
Cost per 100 s.f.	1.15		--		$28.83

Cost of 100 Sq. Ft. (1 Sq.) Glazing and Mottling Over Sand Finish Plaster

0.12 gal. glazing liquid		--	--	$6.60	$ 0.79
Painter	1.60	--	--	24.50	<u>39.20</u>
Cost per 100 s.f.	1.60		--		$39.99

Cost of 100 Sq. Ft. (1 Sq.) Glazing and Highlighting Textured Plaster

0.12 gal. glazing liquid		--	--	$6.50	$0.78
Painter	1.15	--	--	24.50	28.18
Cost per 100 s.f.	1.15		--		$28.96

Cost of 100 Sq. Ft. (1 Sq.) 1-Coat Latex Base Paint Applied to Smooth Plaster

0.31 gal. latex paint		--	--	$14.50	$ 4.50
Painter	0.67	--	--	24.50	16.42
Cost per 100 s.f.	0.67		--		$20.92

Cost of 100 Sq. Ft. (1 Sq.) 1-Coat Latex Base Paint Applied to Sand Finish Plaster

0.37 gal. latex paint		--	--	$14.50	$ 5.37
Painter	0.83	--	--	24.50	20.34
Cost per 100 s.f.	0.83		--		$25.71

Cost of 100 Sq. Ft. (1 Sq.) Starch and Brush Stipple Over Painted Glazed Surfaces

Painter	0.83	--	--	24.50	20.34

Colors in Oil

Colors ground fine in pure linseed oil. One-half pint cans.

Blacks	Price
Drop Black	$0.80
Lamp Black	1.40
Blue	
Palco Blue	3.20
Prussian Blue	2.60
Cobalt Blue	2.60
Ultramarine Blue	2.60

PAINTING AND WALLPAPER

Brown
Raw Turkey Umber 1.20
Burnt Turkey Umber 1.20
Raw Italian Sienna 1.30
Burnt Italian Sienna 1.30
Vandyke Brown 1.20

Green
Palco Green 3.50
Chrome Green, L. 2.10
Chrome Green, M 2.10
Chrome Green, D 2.10

Red
Vermilion 2.70
Rose Lake 2.20
Venetian Red 1.80

Yellow
Chrome Yellow, L. 3.20
Chrome Yellow, M 3.20
Chrome Yellow, O 3.20
Ochre 3.20

Shellac and Varnish

Kind of Material	**per gal.**
Shellac, White, 4-lb. Cut	$ 18.00-22.00
Shellac, Orange, 4-lb. Cut	16.00-20.00
Cabinet Finish Varnish	20.00-26.00
Floor Varnish	19.00-25.00
Flat Varnish	19.00-25.00
Spar Varnish	19.00-29.00
Varnish Remover, Liquid	15.00-25.00
Varnish Remover, Paste	15.00-29.00
Wallpaper Lacquer	12.00-14.00
Wallpaper Lacquer Thinner	8.00-10.00
Lacquer	20.00-25.00
Lacquer Thinner	10.00-12.00

Paints

Wall and ceiling	$14.95
Interior and Exterior Enamel	18.77
House Paint	17.55
Primer (regular)	16.45
Primer (first quality)	17.45
Overcoat/Acrylic Latex (to cover old paint, hardboard or metal)	17.45
Floor and Deck	15.50
Stains and Wood Preservatives (siding, fences, decks)	
Clear Preservative	14.99
Green Preservative	16.99
Semi-Transparent	17.99
Solid Colors	18.99

PAINTING AND WALLPAPER

Labor and Material Required for Interior Painting and Finishing

Finishing Interior Trim—Residential

Description of Work	Painter No. Sq. Ft. per Hr.	Painter Hours 100 Sq. Ft.	Material Coverage Sq. Ft. per Gal.
Preparatory Work for Painting............Sanding	290-300	0.35	....
Sanding and Puttying	190-200	0.50	....
Light Sanding	340-350	0.30	....
Back Priming Interior Trim Up to 6" Wide..........Lin. Ft.*	475-500*	0.21*	1,100-1,200*
Doors and Windows, Painting Interior..........First Coat	140-150	0.70	575-600
Second Coat	115-125	0.85	475-500
Third Coat	115-125	0.85	500-550
Oiling or Priming Wood Sash..........No. of Sash	9-10	0.10	600-700
Base, Chair Rail, Picture Mold, and Other Trim Up to 6" Wide. All Quantities are in Lineal Feet..........First Coat	250-260*	0.40*	1,100-1,200*
Second Coat	165-175*	0.60*	800-900*
Third Coat	140-150*	0.70*	800-900*
Preparatory Work for Enamel Finishes..........Sanding	290-300	0.35	....
Sanding and Puttying	120-130	0.80	....
Light Sanding	135-145	0.72	....
Doors and Windows, Enamel Finish..........Paint First Coat	140-150	0.70	575-600
Undercoat Second Coat	90-100	1.10	375-400
Enamel Third Coat	100-110	1.00	475-500
Four Coat Work. Add ½ Underc. ½ Enamel..........Add'l Coat	90-100	1.10	425-450
Base, Chair Rail, Picture Mold, and Other Trim Up to 6" Wide. All Quantities are in Lineal Feet..........First Coat	250-260*	0.40*	1,100-1,200
Second Coat	145-155*	0.67*	775-800*
Third Coat	125-135*	0.77*	775-800*
Four Coat Work, Add ½ Underc. ½ Enamel..........Add'l Coat	135-145*	0.72*	775-800*
Stain Interior Woodwork..........One Coat	215-225*	0.45	700-725
Stain and Fill Interior Woodwork, Wipe Off..........One Coat	65-70	1.50	400-450
Shellac Interior Woodwork..........One Coat	215-225	0.45	700-725
Varnish, Gloss, Interior Woodwork..........One Coat	165-175	0.60	425-450
Varnish, Flat, Interior Woodwork..........One Coat	170-180	0.57	600-625
Wax and Polish—Standing Trim	100-110	0.95	550-600
Penetrating Stainwax—Standing Trim..........First Coat	170-180	0.57	525-550
Second Coat	190-200	0.51	600-625
Polishing Second Coat	190-200	0.51	....

*Lineal Feet

Labor and Material Required For Interior Painting and Finishing—Con't.

Finishing Interior Trim—Residential

Description of Work		Painter No. Sq. Ft. per Hr.	Painter Hours 100 Sq. Ft.	Material Coverage Sq. Ft. per Gal.
Sanding for Extra Fine Varnish Finish	Sanding	40-45	2.25	550-600
Synthetic Resin Finish, Requires Wiping	First Coat	190-200	0.50	625-675
	Second Coat	210-220	0.47	140-150*
Spackling or Swedish Putty over Flat Trim	One Coat	60-65	1.60	1,050-1,100
Glazing and Wiping over Enamel Trim	One Coat	60-65	1.60	
Brush Stippling Interior Trim, Painted	Varnish	85-90	1.15	600-625
Flat Varnishing and Brush Stippling over Glazed Trim	Stipple	240-250	0.42	

Old Work

Description of Work		Painter No. Sq. Ft. per Hr.	Painter Hours 100 Sq. Ft.	Material Coverage Sq. Ft. per Gal.
Washing Average Enamel Finish	Washing	85-90	1.15	
Washing Better Grade Enamel Finish	Washing	60-65	1.60	
Polishing Better Grade Enamel Finish	Polish	170-180	0.57	1,700-1,800
Removing Varnish with Liquid Remover	Flat Surfaces	30-35	3.00	150-180
Wash, Touch Up, One Coat Varnish	Wash-Touch Up	165-175	0.60	575-600
	Varnish	100-110	0.95	
Wash, Touch Up, One Coat Enamel	Wash-Touch Up	140-150	0.70	
	Enamel	70-75	1.40	375-400
Wash, Touch Up, One Coat Undercoat and	Wash-Touch Up	140-150	0.70	
One Coat Enamel	Undercoat	75-85	1.25	475-500
	Enamel	75-85	1.25	475-500
Burning Off Interior Trim		20-25	4.00	
Burning Off Plain Surfaces		30-35	3.00	

On most commercial and industrial work, a painter will perform 10 to 15 percent more work than given in the above table.
*Per Lb.

PAINTING AND WALLPAPER

Labor and Material Required For Interior Floors

Description of Work		Painter No. Sq. Ft. per Hr.	Painter Hours 100 Sq. Ft.	Material Coverage Sq. Ft. per Gal.
Painting Wood Floors	First Coat	290-300	0.35	525-550
	Second Coat	260-270	0.38	425-450
Filling Wood Floors, Wiping	Fill-Wipe	170-180	0.57	425-450
Penetrating Stainwax—Hardwood Floors	First Coat	390-400	0.25	525-550
	Second Coat	480-490	0.21	600-625
Floor Seal	First Coat	490-500	0.20	600-625
	Second Coat	590-600	0.17	1,200-1,250
Shellac	First Coat	390-400	0.25	500-550
	Second Coat	475-500	0.21	650-675
Stainfill, 1 Shellac, 1 Varnish	Stainfill	260-270	0.38	475-500
	Shellac	390-400	0.25	650-675
	Varnish	300-310	0.33	500-550
Stainfill, 1 Shellac, Wax and Polish	Stainfill	260-270	0.38	475-500
	Shellac	390-400	0.25	650-675
	Wax-Polish	200-210	0.50	1,050-1,075
Varnish, each Coat over Shellac	Varnish	300-310	0.33	500-550
Buffing Floors—by Machine	Buffing	390-400	0.25	
Waxing over 2 Coats of Seal and Polish	Wax-Polish	200-210	0.50	1,050-1,075
Waxing Linoleum—Same as Above				
Linoleum—Varnishing	Varnish	300-310	0.33	500-550
Old Work				
Removing Varnish with Liquid Remover	Flat Surfaces	40-45	2.20	
Clean, Touch Up and Varnish	Clean-Touch Up	125-135	0.77	170-180
	Varnish	300-310	0.33	525-550
Clean, Touch Up, Wax and Polish	Clean-Touch Up	125-135	0.77	
	Wax-Polish	200-210	0.50	1,050-1,075

Labor and Material Required For Painting Interior Walls

Description of Work		Painter No. Sq. Ft. per Hr.	Painter Hours 100 Sq. Ft.	Material Coverage Sq. Ft. per Gal.
Taping, Beading, Spotting Nail Heads and Sanding Gypsum Wallboard	One Coat	90-110	1.00	8-10*
Texture Over Gypsum Wallboard	One Coat	250-270	0.39	325-350
Casein or Resin Emulsion over Textured Gypsum Wallboard	One Coat	300-320	0.32	600-700
Sizing New Smooth Finish Walls	Sizing	350-400	0.27	575-625
Wall Sealer or Primer on Smooth Walls	Sealer	200-225	0.45	275-300
Wall Sealer or Primer on Sand Finish Walls	Sealer	125-135	0.80	575-625
Smooth Finish Plaster—Flat Finish	Sealer	200-225	0.45	500-550
	Second Coat	165-175	0.60	500-550
	Third Coat	175-185	0.55	575-625
	Stippling	190-200	0.52	
Sand Finish Plaster, Flat Finish	Sealer	125-135	0.80	275-300
	Second Coat	135-145	0.70	375-400
	Third Coat	125-135	0.80	400-450
Smooth Finish Plaster, Gloss or Semi-Gloss	Sealer	200-225	0.45	575-625
	Second Coat	170-180	0.57	500-550
	Third Coat	140-150	0.70	500-550
	Stippling	85-95	1.10	
Sand Finish Plaster, Gloss or Semi-Gloss	Sealer	125-135	0.80	275-300
	Second Coat	115-125	0.83	375-400
	Third Coat	125-135	0.80	400-450
Texture Plaster, Average, Semi-Gloss	Sealer	100-110	0.95	250-275
	Second Coat	115-125	0.83	325-350
	Third Coat	125-135	0.80	375-400
Smooth Finish Plaster—Latex Rubber Paints	First Coat	140-160	0.67	300-350
	Second Coat	140-160	0.67	300-350
Sand Finish or Average Texture Plaster—Latex Rubber Paints	First Coat	110-130	0.83	250-300
	Second Coat	110-130	0.83	250-300
Glazing and Mottling over Smooth Finish Plaster		80-90	1.15	1,050-1,075
Glazing and Mottling over Sand Finish Plaster		60-65	1.60	875-900
Glazing and Highlighting Textured Plaster		80-90	1.15	825-850
Starch and Brush Stipple over Painted-Glazed Surface		115-125	0.83	

* Per Lb.

PAINTING AND WALLPAPER

Labor and Material Required For Painting Interior Walls—Con't.

Description of Work	Painter No. Sq. Ft. per Hr.	Painter Hours 100 Sq. Ft.	Material Coverage Sq. Ft. per Gal.
Flat Varnish and Brush Stipple over Painted-Glazed Surface	105-115	0.90	500-550
Texture Oil Paint over Smooth Finish Plaster...............Size	215-225	0.45	675-725
....Texture	40-45	2.30	125-150
Water Texture over Smooth Finish Plaster.......................Size	215-225	0.45	700-725
....Texture	50-55	1.90	85-95
Latex Base Paint, New Smooth Plaster.................First Coat	190-200	0.50	500-550
....Second Coat	210-220	0.47	650-700
On Rough Sand Finish Plaster..................................One Coat	150-160	0.65	325-350
On Cement Blocks...One Coat	135-145	0.70	300-325
On Acoustical Surfaces...One Coat	135-145	0.70	200-225
Casein Paint, over Smooth Finish Plaster...............First Coat	270-280	0.37	500-550
....Second Coat	300-310	0.33	500-550
Over Rough Sand Finish Plaster..............................One Coat	250-260	0.40	325-350
Over Cement Blocks..One Coat	165-175	0.60	300-325
Over Acoustical Surfaces...One Coat	165-175	0.60	200-225
Over Cinder Concrete Blocks...................................One Coat	125-135	0.77	125-150.

When applying paint to walls and ceilings using a roller applicator, increase above quantities 10 to 15 percent and reduce painter time by the same amount.

Labor and Material Required For Painting Interior Walls

Old Work Description of Work	Painter No. Sq. Ft. per Hr.	Painter Hours 100 Sq. Ft.	Material Coverage Sq. Ft. per Gal.
Washing off Calcimine—Average Surfaces	115-125	0.83	
Washing Smooth Finish Plaster Walls—Average	145-155	0.67	
Washing Sand Finish Plaster Walls—Average	100-110	1.00	
Washing Starched Surfaces and Restarching, Smooth Surfaces......Washing	145-155	0.67	
Restarching	130-140	0.75	
Removing Old Wall Paper—Not Over 3 Layers	65-75	1.45	
Washing Off Glue After Removing Paper (including Fixing Average Cracks and Sizing)	125-135	0.77	
Cutting Hard Oil or Varnish Size Walls (including Fixing Average Cracks)	125-135	0.77	
Cutting Gloss Painted Walls (including Fixing Cracks)	125-135	0.77	
Washing, Touch Up, One Coat Gloss Paint to Smooth Plaster Surfaces......Wash-Touch Up	125-135	0.77	
One Coat	125-135	0.77	450-500
Synthetic-Resin Emulsion Paint over Old Painted Walls...........One Coat	190-200	0.50	400-425
Wallpaper, Canvas, Coated Fabrics, Paper Hanging, Wood Veneer			
Canvas Sheeting	50-60	1.82	
Coated Fabrics.............................Single Roll	2-2½		
Wallpaper—One Edge Work..................Single Roll	3½-4		
Wire Edge Work..........................Single Roll	2½-3		
Butt Work................................Single Roll	2¼-2¾		
Scenic Paper 40″x60″.......................Single Roll	1-1¼		
Wood Veneer......................................Sq. Ft.	12-13		
Lacquer Finish Over Wood Veneer...................First Coat	300-310	0.33	275-325
Second Coat	300-310	0.33	450-500
Penetrating Wax or Synthetic Resin Application over Wood Veneer......First Coat	550-575	0.18	550-600
Second Coat	600-625	0.1	650-675

PAINTING AND WALLPAPER

WALLPAPER

Wallpaper is estimated by the roll. Standard rolls are 18" or 20-1/2" wide and eight yards long, containing 36 sq. ft. or 4 sq. yds. There is a double roll that contains twice as much as the single roll: 72 sq. ft. or 8 sq. yds. of paper.

Estimating the Quantity of Wallpaper Required for any Room. Measure the entire distance around the room in lin. ft. and multiply by the height of the walls or the distance from the floor to the ceiling. The result will be the number of square feet of surface area. Make deductions in full for the area of all openings such as windows, doors, consoles, mantels or built-in bookcases. The difference between the area of the walls and the area of the openings will be the actual number of sq. ft. of surface to be papered.

When adding for waste in matching, cutting and fitting, paper hangers use different methods and allow different percentages, contingent upon the height of the ceiling, the design of the paper, and the size of the patterns or figures, but an allowance of 15% to 20% should be sufficient in nearly all cases.

Bear in mind that when a border is used, it is not necessary to run the side wallpaper more than one or two inches above the bottom of the border. The height of the wood base should also be deducted from the total height of the wall.

In rooms with drop ceilings, the depth of the drop should be deducted from the height of the wall, which will decrease the area of side wallpaper in the same amount as it increases the quantity of ceiling paper.

If a border is required, measure the distance around the room and the result will be the number of lin. ft. of border. Dividing by 3, gives the number of yards of border required.

When estimating the quantity of wallpaper required for ceilings, multiply the width of the room by the length, and the result will be the number of sq. ft. of ceiling to be papered. In rooms with a drop ceiling, add the depth of the drop, or the distance the ceiling paper extends down the side walls, to the length and breadth of the room and multiply as described above.

After the side walls and ceiling area have been computed in sq. ft., divide by 36. The result will be the number of single rolls of paper required; or dividing by 72 gives the number of double rolls of paper required.

Paste Required Hanging Wallpaper. The quantity of paste required for hanging wallpaper will vary with the weight of the paper and the surface to which it is applied. Where light- or medium-weight wallpaper is used, one gallon of paste should hang 12 single rolls of paper.

If heavy or rough texture paper is used, it will often be necessary to give it two or three applications of paste to obtain satisfactory results. On work of this kind one gallon of paste should hang 4 to 6 single rolls of paper. There are prepared pastes on the market that require only the addition of cold water to make them ready to use.

One lb. of prepared cold water dry paste should make 1-1/2 to 2 gallons of ready-to-use paste. Prepared cold water dry paste to make 3 gallons of ready-to-use paste should cost as follows:

	Hours	Rate	Total	Rate	Total
2 lbs. dry paste	--	--	--	$ 0.25	$0.50
Laborer	0.25	--	--	24.35	6.09
Total cost 3 gals. paste	0.25	--	--		$6.59
Cost per gal.			--		2.20
Cost per roll lgt. wt. paper			--		0.18
Cost per roll hvy. wt. paper			--		0.42

Labor Hanging Wallpaper. Different methods of hanging wallpaper are used in different sections of the country. In some localities a paper hanger and helper work together, one trimming the paper and pasting, the other hanging it. However, in many cities the work is performed by one worker who cuts, trims, fits, pastes and hangs the paper.

The quantity and costs given on the following pages are based on the performance of one worker but will prove a fair average for use in any locality.

Hanging Wallpaper on Walls and Ceilings, One Edge Work. When hanging light- or medium-weight paper on ceilings and drops, a paper hanger should hang 28 to 32 single rolls of paper per 8-hr. day at the following labor cost per roll:

	Hours	Rate	Total	Rate	Total
Paper hanger	0.27	--	--	24.35	$6.57

Hanging Wallpaper on Walls, Butt Work. Where light- or medium-weight paper is used for bedrooms and halls, and where a good grade of workmanship is required with all paper trimmed on both edges and hung with butt joints, a paper hanger should trim, fit and hang 20 to 24 single rolls of paper per 8-hour day at the following labor cost per roll:

PAINTING AND WALLPAPER

	Hours	Rate	Total	Rate	Total
Paper hanger	0.36	--	--	24.35	$8.77

One thing that will slow up work is where new paper is applied over old rough textured wallpaper such as oatmeal designs.

Hanging Wallpaper, First Grade Workmanship. Where a good grade of medium- or heavy-weight wallpaper is used, with all paper hung with butt joints, a paper hanger should trim, fit and hang 16 to 20 single rolls of paper per 8-hour day at the following labor cost per roll:

	Hours	Rate	Total	Rate	Total
Paper hanger	0.44	--	--	24.35	$10.71

Labor Hanging Scenic Paper. Where scenic paper or paper with mural designs is used, a paper hanger will hang only 8 to 10 single rolls per 8-hour day at the following labor cost per roll:

	Hours	Rate	Total	Rate	Total
Paper hanger	0.9	--	--	24.35	$21.92

Placing Wallpaper Borders. Wallpaper borders are usually estimated by the lin. ft. or lin. yd., and as the width varies from 3" to 18", there is never over one width required. The average room requires 50 to 60 lin. ft. or 17 to 20 yds. of border.

A paper hanger should place border in a room of this size in 30 to 40 minutes, at a rate of 100 to 125 lin. ft. per hour. The labor cost per 100 lin. ft. (33-1/3 yds.) should run as follows:

	Hours	Rate	Total	Rate	Total
Paper hanger (per 100 l.f.)	0.9	--	--	24.35	$21.92
Cost per l.f.			--		0.22
Cost per yd.			--		0.66

Hanging Coated Fabrics. Where coated fabrics such as Walltex and Sanitas are used, a paper hanger should hang 16 to 20 single rolls per 8-hour day at the following labor cost per roll:

	Hours	Rate	Total	Rate	Total
Paper hanger	0.44	--	--	24.35	$10.71

Canvassing Plastered Walls. When applying canvas to plastered walls, an average mechanic should apply 50 to 60 sq. ft. per hour at the following cost per 100 sq. ft.:

	Hours	Rate	Total	Rate	Total
12 sq. yds. canvas		--	--	$2.00	$24.00
4 lbs. dry paste		--	--	0.25	1.00
Paper hanger	1.82	--	--	24.35	44.32
Cost per 100 s.f.			--		$69.32
Cost per s.f.			--		0.69

Flexwood. Flexwood is a genuine wood veneer cut to 1/85", glued under heat and hydraulic pressure to cotton sheeting with a waterproof adhesive. A patented flexing operation breaks the cellular unity of the wood to produce a limp, pliable sheet that may be applied by hand to any smooth surface, flat or curved. Waterproof Flexwood 710 Adhesive, which makes a permanent bond, is used to apply Flexwood. Standard sizes of stock material are 18" and 24" widths, and 8'-0" and 12'-0" lengths.

The mechanics of installing Flexwood are as follows. The background is sized with Flexwood cement. Another coating of cement is brushed on the Flexwood, which is then hung in the manner of any sheet wall covering. A stiff, broad knife used with considerable pressure smoothes out the Flexwood, removes air spaces and furnishes the necessary contact.

Flexwood is available in woods such as mahogany, walnut, oak, prima vera, knotty pine, orientalwood, satinwood, zebrawood, rosewood, English oak, maple and lacewood. The cost of the material varies from $1.20 to 3.75 per sq. ft., depending on the kind of wood. Most woods run from $1.50 to 1.75. Adhesive runs $9.50 per gallon.

It requires a skilled and thoroughly competent paper hanger to apply this material and the labor costs vary considerably with the design to be obtained. Considerable time is usually required to lay out any room to obtain the desired effect and spacing of strips. This is especially true where certain designs must be obtained on columns, walls and pilasters and where narrow strips of contrasting wood are used to obtain inlay effects.

PAINTING AND WALLPAPER

On plain walls without inlays, a paper hanger experienced in this class of work should apply 95 to 105 sq. ft. of Flexwood per 8-hour day at the following labor cost per 100 sq. ft.:

	Hours	Rate	Total	Rate	Total
Paper hanger	8.0	--	--	$24.35	$194.80
Cost per s.f.			--		1.95

On pilasters, columns and walls inlaid with narrow strips of contrasting woods, the work is considerably slower due to the additional cutting and fitting required. On work of this kind, a paper hanger experienced in this class of work should apply 50 sq. ft. per 8-hour day at the following labor cost per 100 sq. ft.:

	Hours	Rate	Total	Rate	Total
Paper hanger	16.0	--	--	$24.35	$389.60
Cost per s.f.			--		3.90

Additional time for paper hanger or foreman will be required laying out the work into patterns, marking off strips, etc. This will vary with the size of the job. Add for overhead expense and profit. After the Flexwood is applied, the joints should be sanded lightly to remove any imperfections.

Finishing Flexwood. Flexwood will take any wood finish, but where the natural color of the wood is desired, the Flexwood is given one coat of lacquer sealer, sanded lightly between coats and then given one coat of lacquer for a finish coat. One gallon of lacquer sealer will cover about 300 sq. ft. of surface, one coat. After the sealer has been applied, one gallon of lacquer should cover 450 to 550 sq. ft.

An experienced painter should apply lacquer sealer, sand lightly between coats and apply one coat of lacquer to 90 to 110 sq. ft. of surface per hour at the following labor cost per 100 sq. ft.:

	Hours	Rate	Total	Rate	Total
Paper hanger	1.0	--	--	24.35	$24.35
Cost per s.f.			--		0.24

Vinyl Wall Covering. Vinyl wall covering is composed of a woven cotton fabric to which a compound of vinyl resin, pigment and plasticizer is electronically fused to one side. It comes 54" wide in 30-yard rolls in three weights: heavy (36 oz. per lin. yd.); medium (24-33 oz. per lin. yd.); and light (22-24 oz. per lin yd.). It is available in more than 1,000 color pattern combinations.

Some of the patterns available simulate linens, silks, moires, grasses, tweeds, Honduras mahogany, travertine, damasks, and burlaps. Costs range from $0.35 to 0.90 per sq. ft.

Vinyl fabric is hung using regular wallpaper hanging procedures for hanging fabric baked wallcoverings. A broad knife is used to smooth out any wrinkles or air pockets and insures a good contact. Wash off any excess paste that remains on the surface of the material. No further finishing is necessary.

A competent paper hanger can apply approximately 400 sq. ft. per 8-hour day at the following costs per 100 sq. ft.:

	Hours	Rate	Total	Rate	Total
Paper hanger	2.0	--	--	24.35	$48.70
Cost per s.f.			--		0.49

One gallon of paste covers approximately 10 lin. yds. of 54" width material. To the above add overhead expenses and profit. If excess cutting and fitting is required, the above figures should be increased to allow for the above conditions.

Chapter 9

PLUMBING

Preparation of an accurate, detailed estimate on plumbing and sewerage requires a working knowledge of the trade. All work must be laid out on paper and the number of lineal feet of each kind of pipe—tile or cast iron sewers, soil and vent pipe, water pipe, gas pipe, drains, valves and pipe fittings—must be estimated separately. Some general contractors and estimators possess this knowledge, but many do not, and it is difficult for the latter to prepare more than an approximate estimate on the various kinds of plumbing work.

Budget figures for average plumbing installations can be estimated in one of two ways: as a percentage of the total cost of the project, or as the grand total of a predetermined cost allowance for the installation of each fixture. Plumbing estimates as a percentage of the total job will vary from a minimum of 3% on up to 12%, depending on the number and type of fixtures required and on their locations in the building. Percentage estimates are valid for general plumbing only. Process and other special condition piping would have to be entered as individual items on an estimate.

When figuring plumbing work where jobs are less than 4 hours, multiply all hourly rates by 1.9 for service work and 1.8 for remodeling work. The private residence is difficult to peg. Where kitchens and baths can be back to back, the percentage is on the low side. But in a large, rambling house with a central powder room area with kitchens, family and master bedrooms all in separate wings and with hose bibs and lawn sprinkling systems on all four sides and sunken baths, shell-shaped lavatories, and gold plated fixtures, even 12% may be too low.

The allowance per fixture estimate is generally more accurate. An experienced plumber can develop accurate prices on this basis, though he will have to use his judgement on how much to add for bringing in the main waste and supply lines, which depends on how spread out the project is.

The *per fixture price* is the cost of the fixture itself plus the cost of the immediate piping connections and installation within the room. Some per fixture costs have added to them a prorated cost of the central piping system. It would be more accurate to add this cost into the cost of bringing waste and supply lines from the street to the building and permit costs and overhead and profit, all as separate items, because some buildings require a much more extensive central system to supply the same or even fewer fixtures than others. Per fixture allowances are discussed later in this chapter.

PLUMBING SYSTEMS

Estimating Quantities and Costs of Sewer Work. Sewerage is estimated by the lineal foot, obtained by measuring the number of lineal feet of each size sewer pipe required, such as 4", 6", 8" and 12" pipe. All fittings should be listed in detail, such as the number of ells, Y's and T's. This method is used both for cast iron and tile sewer pipe.

Most plumbers estimate their sewer work at a certain price per lineal foot, which includes excavating, sewer pipe and labor. This is all right on ordinary jobs, but if there is an exceptionally large amount of deep excavation (from 7 to 15 feet deep), then the excavating, shoring and bracing should be estimated separately.

The best method is to estimate the number of cubic yards of excavating based on a trench 1'-6" wider than the pipe diameter. Where 4" to 6" sewer pipe is used, an experienced sewer layer should lay 12 to 15 lin. ft. of pipe per hour, or 100 to 125 lin. ft. per 8-hour day. Add for excavating and additional time for setting catch basins, gravel basins and triple basins.

The cost of excavating a trench 3'-0" deep and placing 100 lin. ft. of 6" tile sewer pipe should be estimated as follows:

	Hours	Rate	Total	Rate	Total
Labor excavating	39.0	--	--	$21.50	$ 838.50
Labor backfilling	10.5	--	--	21.50	225.75
100 lin. ft. 6" pipe		--	--	5.25	525.00
Pipe layer	7.0	--	--	29.30	205.10
Total cost per 100 l.f.	56.5		--		$1,794.35
Total per l.f.			--		17.94

Estimating Quantities of Cast Iron Soil Pipe, Downspouts, Stacks and Vents. When estimating soil pipe, stacks and vents, note the number of stacks and length of each. List the number of lineal feet of each size pipe; the number of pieces, size and kind of fittings required; and the price at current market prices. The labor cost of placing cast iron pipe is given on the following pages.

Estimating Quantity of Pipe and Fittings. When estimating the quantity of black or galvanized pipe, list the number of lineal feet of each size pipe required, such as 1/2", 3/4", 1", 1-1/4", 1-1/2", 2", 2-1/2" or 3", and estimate at the current market price. After the cost of the pipe has been computed, take 60% to

75% of the cost of the pipe to cover the cost of all fittings required. Brass and copper pipe and fittings should be estimated in the same manner as galvanized pipe.

Estimating the Quantity and Prices of Valves. All valves of the different sizes should be listed on the estimate separately, stating the number of each size required and pricing them at current market price.

Estimating the Quantity of Fixtures. Each type of fixture such as sinks, lavatories, laundry trays, water closets, bathtubs, drinking fountains, shower baths, hot water tanks and heaters, house pumps and bilge pumps should be listed separately and priced at current market prices.

Labor Roughing-in for Plumbing. The normal practice is to allow a certain percentage of the cost of the roughing materials to cover the labor cost of installing them. This will vary with the type of building and grade of work.

The total cost of roughing-in materials is computed. This means all materials required for roughing in the job—cast iron soil pipe; downspouts, stacks, vents and fittings; all black and galvanized pipe and fittings; all valves, increasers, tees, Ys, 1/8 bends and nipples. Then, the labor cost is estimated at a certain percentage of the material cost.

In medium priced one- and two-story residences, the labor cost of roughing-in varies from 75% to 85% of the cost of the roughing materials, with 80% a fair average.

In apartment buildings of non-fireproof construction, the labor cost of roughing-in averages about 90% to 100% of the cost of the roughing-in materials, while on high grade fireproof construction, the labor cost is from 100% to 120% of the cost of materials. Sewer work is estimated separately from roughing-in.

Labor Handling and Placing Plumbing Fixtures. The labor cost handling and placing all kinds of fixtures, such as laundry tubs, kitchen sinks, lavatories, bathtubs, shower tubs, water closets and drinking fountains, is usually estimated at 25% to 30% of the cost of the fixtures.

Overhead and Profit. Plumbing contractors like to add 10% to the cost of their work for overhead expense and another 10% for profit, but frequently jobs are let on the basis of a straight 10% or less, depending on the competition and how badly the contractor needs the work.

PREPARING DETAILED PLUMBING ESTIMATES

For a detailed plumbing estimate, you must list separately the quantity for each class of work, such as the number of lineal feet of 4", 6" and 8" sewers and fittings; the number of lineal feet of each size of soil pipe, vents and fittings; the number of lineal feet of water pipe and fittings of different materials and sizes; the number of valves of the various kinds and sizes; the number of each type of fixture required, such as hot water heaters, laundry trays, kitchen sinks, slop sinks, bathtubs, lavatories, water closets and shower baths; and a list of fittings and supplies required for each.

The following is a list of items the plumbing contractor should include in his estimate:

1. Sewerage, including double sewer system.
2. House tanks, foundations, etc.
3. Compression tanks, foundations, etc.
4. Sewer ejector and bilge pump and basins.
5. Iron, gravel, catch and other basins.
6. Water filters, foundations.
7. Water meter.
8. Soil and vent pipe and fittings.
9. Roof flashings.
10. Shower traps.
11. Closet bends.
12. Drum traps.
13. Lead, solder, sundries.
14. Inside downspouts.
15. Outside downspouts.
16. Downspout heads.
17. Iron sewer and fittings.
18. Floor drains.
19. Back water gates.
20. Service pipe to building.
21. Mason hydrant.
22. Stop cock and box.
23. Galvanized pipe and fittings.
24. Brass or copper pipe and fittings.
25. Valves, check valves.
26. Pet cocks, sill cocks, hose bibbs.
27. Hot water tank and heater.
28. Pipe covering, tank covering, heater covering, filter covering, painting.
29. Gas fitting, mains ranges, heaters, etc.
30. Manholes and covers.
31. Catch basins and covers.

PLUMBING

32. Surface drain basins and covers.
33. Fire system, including pumps, hose, Siamese connection, meter, etc.
34. Roughing-in material.
35. Fixtures, such as w.c., lavatories, bathtubs, shower baths, etc.
36. Permits, insurance, trucking, freight, telephone, watchman, etc.
37. Overhead expense.
38. Profit.

PIPE AND PIPE FITTINGS

All prices on all types of pipe and pipe fittings and supplies should be checked with a supplier at the time of the job, because they are subject to continual change.

Approximate Lineal Foot Price of Cast Iron Soil Pipe

Size	Single Hub		Double Hub	
	Service	Extra Heavy	Service	Extra Heavy
2"	$3.10	$4.30	$4.25	$5.85
3"	4.00	5.05	4.94	6.85
4"	4.75	6.75	6.60	8.98
5"	6.20	8.65	8.40	11.80
6"	7.30	10.10	9.80	13.80
8"	11.25	15.70	15.40	21.35
10"	18.00	25.30	24.70	33.70
12"	25.85	36.30	35.95	48.90

Prices of Cast Iron Soil Pipe Fittings

Description of Fitting	Sizes in Inches					
	2"	3"	4"	5"	6"	8"
Quarter bends, service	$3.40	$6.20	$9.00	$11.80	$15.70	$51.70
Quarter bends, extra heavy	3.90	7.30	10.30	14.00	17.60	58.40
Eighth bends, service	2.50	5.05	7.30	10.30	12.35	35.95
Eighth bends, extra heavy	2.80	5.70	8.40	11.80	14.00	41.60
Sanitary T&Y branches, extra heavy	7.00	13.50	16.85	24.70	38.20	78.65
Plain traps S P or 3/4, service	9.60	12.90	18.00	34.80	44.90	102.00
T cleanout, extra heavy	30.30	40.45	57.30	103.40	148.30	185.40

Comb. Y & eighth bend,
 extra heavy.............................. 10.10   13.50   19.10   37.10   45.00 104.50

Closet bends slip collr type—4" x 4" x 12"—$23.00; 4" x 4" x 18"—$27.00

Lineal Foot Prices of Buttweld Steel Pipe

Size	Standard Weight		Extra Heavy	
	Black	Galvanized	Black	Galvanized
1/4"	$0.33	$0.50	$0.50	$0.67
3/8"	0.45	0.60	0.65	0.70
1/2"	0.47	0.65	0.65	0.80
3/4"	0.67	0.80	0.85	1.00
1"	1.00	1.25	1.30	1.45
1-1/4"	1.20	1.60	1.65	1.90
1-1/2"	1.50	1.85	1.95	2.30
2"	2.00	2.50	2.65	3.10
2-1/2"	2.90	3.65	4.10	4.75
3"	3.80	4.75	5.40	6.30
3-1/2"	4.70	5.85	6.70	7.85
4	5.60	6.85	8.00	9.30

Prices on Standard Malleable Iron Fittings

Fitting	Size								
	1/2"	3/4"	1"	1-1/4"	1-1/2"	2"	3"	4"	6"
Tee - Black	$0.62	$0.90	$1.60	$2.60	$3.15	$4.65	$11.80	$24.15	$67.40
Tee - Galv.	0.90	1.30	2.10	3.70	4.50	6.00	16.30	33.70	95.50
90° El - Black	0.50	0.55	1.05	1.70	2.25	3.20	10.70	18.00	53.90
90° El - Galv.	0.75	0.80	1.50	2.35	3.15	4.50	14.60	25.30	73.00
45° El - Black	0.75	0.95	1.25	2.00	2.50	3.40	11.80	19.10	61.80
45° El - Galv.	1.05	1.25	1.75	2.50	3.40	3.60	16.30	26.60	84.30

PLUMBING

Lineal Foot Prices of Copper Water Tubing

	\multicolumn{10}{c}{Nominal Size}

	1/4"	3/8"	1/2"	5/8"	3/4"	1"	1-1/4"	1-1/2"	2"	3"
Light Type M	0.40	0.45	0.67	0.80	0.95	1.30	1.80	2.35	2.30	6.00
Med. Type L	0.40	0.60	0.85	1.00	1.25	1.70	2.20	2.80	3.75	7.25
Heavy Type K	0.45	0.70	0.95	1.10	1.60	2.00	2.50	3.25	4.90	9.55

Prices of Wrought Copper Soldered Joint Fittings

	\multicolumn{8}{c}{Nominal Size}

Type	1/2"	3/4"	1"	1-1/4"	1-1/2"	2"	3"	4"
Tee	$0.22	$0.60	$1.70	$2.65	$3.37	$5.40	$17.65	$35.95
90° Elbow	0.17	0.38	0.80	1.25	1.70	2.65	9.00	18.00
45° Elbow	0.28	0.50	0.85	1.45	1.90	2.50	10.95	22.50

PIPING SPECIALTIES

Cast Iron Drum Traps

Drum trap with cover and gasket, 4" x 8", 2 inlets.......................... $10.50
Drum trap with cover and gasket, 4" x 8", 3 inlets........................... 12.10

Roof Drains

	3"	4"	5"	6"	8"
Cast Iron	$130.00	$135.00	$195.00	$215.00	$340.00
Galvanized	220.00	225.00	280.00	300.00	425.00

Floor Drains

Type	3" outlet	4" outlet	5" outlet
Flat round top cast iron	$40.00	$50.00	$85.00
Flat square top cast iron	55.00	55.00	95.00
Funnel type brass	80.00	80.00	
Drain with bucket cast iron	110.00	110.00	115.00
Trench drain 10" x 24"	130.00	150.00	

Septic Sewage Disposal Tanks

Septic tanks are used in rural or suburban districts where running water is available but no sewers. The capacity of a septic tank should be at least equal to the maximum daily flow of sewage, normally estimated at 50 gallons per person per day. For part time service in factories, churches and schools, 25 gallons per person per day is generally satisfactory for estimating tank capacities. In some localities health authorities require at least a 500-gallon tank for residential installation.

They are commonly furnished in 12- and 14-gauge copper bearing steel, electrically welded and covered with a thick coating of asphalt to protect the tanks against corrosive action.

Size of Tank	Capacity Gallons	Size Tile Connection	Weight Lbs.	Capacity No. Persons	Price Each
46" dia. x 48"	300	4"-6"	275	5-6	$67.50
52" x 60"	500	4"-6"	380	6-7	102.00
46" x 120"	750	6"	575	7-12	140.00
48" x 144"	1000	6"	980	12-15	195.00

Prices of Brass Valves

Kind of Valve	Size in Inches							
	1/2"	3/4"	1"	1-1/4"	1-1/2"	2"	2-1/2"	3"
Gate valve, 100 lbs. pressure	8.00	10.00	13.00	16.50	19.25	30.00	--	--
Gate valve, 125 lbs. pressure	11.00	13.75	17.00	22.00	28.00	40.00	80.00	120.00
Globe valve, 100 lbs. pressure	9.50	12.50	17.00	22.00	29.00	43.00	--	--
Globe valve, 125 lbs. pressure	10.00	13.50	18.00	25.00	32.00	48.00	80.00	125.00
Check valve, horizontal 125 lbs	15.00	20.00	28.00	36.00	45.00	74.00	100.00	140.00
Check valve, 125 lb. swing	12.00	14.00	18.00	23.00	27.00	39.00	70.00	100.00

Angle valves, same price as Globe valves.

Gas Heaters

Capacity Gals.	Overall Diameter	Overall Height	Approx. Weight	Price 5 Year	Price 10 Year
20	14-1/2"	61"	144 lbs.	$100.00	$120.00
30	16-1/4"	64"	174	120.00	170.00
40	18-1/2"	65"	195	150.00	185.00
50	20-1/4"	65"	274	190.00	240.00

Electric Heaters

Capacity Gals.	Overall Diameter	Overall Height	Approx. Weight	Price 10 Year
40	20	50	160	$170.00
52	23	50	181	200.00
66	23	62	227	220.00
82	25	63	270	250.00
110	28	69	430	290.00

PLUMBING FIXTURES AND TRIM

When plumbing first moved indoors, it was generally installed in a full-size room and was arranged about the walls as so much furniture. Later, sanitary concerns became paramount, and kitchens and baths were designed the same as machines—all chrome and tile and porcelain enamel and reduced to minimum dimensions to facilitate cleaning.

When maids and servants became a thing of the past, the kitchen emerged as an entertainment center where family and friends gather. Today it is often difficult to tell where the family room ends and the kitchen begins. In many upscale residences the bathroom seems to be following suit, especially in the bathing area, which is often a separate room with built-in equipment for exercise and relaxation and sometimes even a sauna, greenhouse and fireplace! Often the contractor will be called in to convert a spare bedroom into a bath and dressing room suite.

The following discussion of fixtures and their prices gives figures that include the fittings. Also included are averages for the piping hook-ups to the main risers and stacks that might be added to arrive at per fixture installed budget figures.

Bathtubs. The least expensive, standard, recessed tub is made of enameled steel. This costs about $300 to $325 for the 4'-6" length and $350 to $400 for the 5'-0" length. Units made of cast iron cost about $450 to $475 in those sizes in white. Add $100 for color. If these units are to be recessed into the floor, add another $20.00. Square recessed tubs 42" x 48" cost $390 in cast iron.

Fiberglass units complete with three enclosing walls are about $275.00. A plumber with a helper should set a tub with shower in about two hours. To arrive at a per-fixture installed budget figure, add about $210.00.

Shower Stalls. These range from simple baked enamel on steel, which cost about $100, to custom marble enclosures costing six times that. Prefabricated fiberglass units are very popular and come with built-in seats and grab bars for the elderly. A 3' x 3' unit costs $235 complete.

Often the plumber will find his work limited to furnishing the receptor and mixing valve and outlet with other trades providing finished walls. A precast receptor pan in terrazzo together with chrome fittings costs $90 in 3' x 3' size and $110 in 3' x 4'. One worker can install it in about four hours.

Lavatories. Wall-hung lavatories are available in porcelain-enameled cast iron or vitreous china. A 19" x 17" unit costs $125.00 in white, $175.00 for color; $100 china white, $120 china colored. A 22" long white unit costs $140 cast iron white, $155.00 cast iron colored and $170 china white and $180.00 china colored.

Allow two workers one hour to install. Add $230.00 to arrive at a per fixture installed price. Drop-in countertop units are usually self-rimming. They come in cast iron, china, steel, stainless and molded plastic. An 18" round unit of cast iron with vitreous enamel costs about $65.00 in white, $75.00 in color; in stainless steel, it is $120.00, and in enameled steel, $50.00. A 20" x 18" unit costs about $10 more. Two workers require 1-1/2 hours to set a unit.

Also available for vanity units are precast, one-piece bowl and countertops in a variety of marbleized colors and patterns. Generally, these are manufactured locally, and prices should be checked, but a 2' long top and bowl should cost about $75.00 and take two workers 1-1/2 hours to set, and a 6' top with double bowl should cost about $175.00 and take 3 hours to set.

Water Closets. Water closets can be either wall-hung or floor-mounted and have a tank or flush valve water supply. The tank can be a separate piece, or the unit can be cast in one piece with the tank dropped behind the seat.

A floor-mounted one-piece will cost $240.00 in white, $285.00 in color. A floor-mounted two- piece unit costs $110 in white, $135 in color. For areas where water supply is a problem, units designed to flush with as little as two quarts are available at about $400. Floor-mounted units will take two workers 1.5 hours to set. For a per fixture installed budget figure, add $180.

Bidets. For the ultimate touch in the luxury bath, install a bidet. It will cost about $250.00 in white, $290.00 in color and take two workers two hours to set.

Sinks. Kitchen sinks are usually self-rimming and set into a cabinet top of plastic or tile. Single-bowl sinks are pretty common these days, because automatic dishwashers have eliminated the need for double-bowls. In luxury kitchens a single bowl might be set where the dishes are washed, with a second single-bowl near the refrigerator for washing vegetables and a third near the ice making machine for making drinks.

A single bowl 24" x 21" costs $75.00 in enameled steel, $112.50 in white cast iron, $137.50 in colored cast iron and $156.25 in stainless. Figure two workers three hours to set. Add $375.00 for a per fixture installed price.

A double unit 42" x 21" costs $156.25 in steel, $205.00 in cast iron white, $225.00 in cast iron color and $250.00 in stainless. Two workers will take 3.5 hours to set such a unit.

Cast iron sink tops with drain board cost about $225.00 for single-bowl 42" unit, $260.00 for double-bowl 60" unit.

Enameled steel sink and drain board tops cost about $150.00 for a 42" length, $200.00 for a double-bowl 60" length.

Laundry sinks in enameled cast iron cost $150.00 for a single compartment, $225.00 for a double; in cast stone, $75.00 and $115.00; in fiberglass, $60.00 and $90.00; in stainless steel, $150.00 and $250.00. Two workers will set a single unit in 1.5 hours; a double in 2 hours.

Electric Dishwashers. Electric dishwashers can be installed as separate units or form part of a complete sink. A dishwasher unit only, with floor cabinet but without top and backsplash, used under a continuous countertop 24" x 25" x 34-1/2" high, costs $400.00.

A dishwasher unit with porcelain enamel top and backsplash, cabinets made of electrically welded rust-resistant steel with baked on enamel finish, is 23" x 27" x 39" high and costs $430.00. Add installation charges to the above prices.

Garbage Disposal Units. These self-contained units are attached to the kitchen sink to form an enlarged drain into which all kitchen food wastes can be placed. The unit consists of a housing in which is contained a propeller and shredding mechanism. A 1/4 h.p. motor is directly connected to the propeller and shredding mechanism and supplies the power for operating the waste unit. The price of this unit averages about $135.00, to which installation charges must be added.

Chapter 10

ELECTRICAL

Electrical work is almost always a part of any major remodeling job. All room additions and conversions will likely involve adding wiring, outlets and fixtures. For remodeling existing kitchens and baths, it quite likely will be necessary to upgrade the electrical to meet current code requirements and accommodate new appliances.

In rehabbing older homes it might be necessary to replace the entire service. If wiring has become worn, or major portions of the system have sustained damage, strong consideration should be given to replacement starting at the service entrance. In some areas, a building permit will not be issued unless older wiring is brought up to present codes.

Whenever remodeling work is planned, it is a good time to assess electrical needs and add new service. The electrical job is simpler and less expensive when it is part of a remodeling job.

In addition to more outlets to meet code, which in some localities calls for an outlet every 12 feet, when major appliances or central air conditioning is added, there will often be a need to provide additional power. Again, the need is more likely to occur in older homes where the service is no longer adequate. It is not unusual to find that the original service is rated at 60 amps, 115 volts. Unless the home is unusually small, a more realistic service for today's needs is 100 to 120 amps, 220 volts. Many large appliances, ranges and air conditioners in particular, require 220 volts.

Whenever a remodeling project includes the addition of a complete room or other new construction, the need for including electrical service is self-evident. Provision for power is part of the basic plan and is ordinarily shown on construction drawings, if they have been prepared. If extensive drawings are not part of the project, be ready to suggest to the customer where service ought to be provided to satisfy any present or future need. New construction presents a unique opportunity to add all the convenience desired, at a time when the wiring can be done very easily as part of the total job.

Opportunities also exist when the project is remodeling or finishing a previously unfinished area, such as a basement or an attic. As walls are roughed in but before any finishing work is begun, wiring can be done in a minimum of time, without having to disturb existing construction. It is a good time, too, to suggest further improvements to the customer, making the end result of the project even more pleasing and convenient.

Finally, there is the question of codes. Every locality has construction codes that cover electrical service. These are, of course, designed to insure that safe wiring is provided and that the homeowner gets what he pays for. Since a building permit is frequently required for major remodeling jobs, the work will be inspected. It's important that electrical service be brought up to date, in accordance with local codes, so that the inspection will be successful. It's also important to consult local codes at the time a job is being planned and estimated.

Special Problems to Consider. Electrical service in remodeling generally requires opening or removing portions of walls, ceilings, and even floors, and it is often necessary to remove old wiring. This is especially true when rigid conduit was used originally, and the plans call for using some or all of what is there. If wiring must be upgraded, the old wires must be removed, unless the conduit is of adequate size to accommodate the new along with the old. The job may be as simple as disconnecting the old and trimming the ends, or as complicated as pulling out every length of existing wire for reasons of safety, space or to meet code requirements. Use the following to determine the capacity of standard sizes of conduit.

Sizes of Conduit

Size of Wires	Number of Wires				
	2	3	4	5	6
TW					
14	1/2"	1/2"	1/2"	1/2"	1/2"
12	1/2"	1/2"	1/2"	1/2"	1/2"
10	1/2"	1/2"	1/2"	1/2"	3/4"
8	1/2"	3/4"	3/4"	1"	1"
THW					
14	1/2"	1/2"	1/2"	1/2"	1/2"
12	1/2"	1/2"	1/2"	3/4"	3/4"
10	1/2"	1/2"	1/2"	3/4"	3/4"
8	3/4"	3/4"	1"	1"	1-1/4"
TW and THW					
6	3/4"	1"	1"	1-1/4"	1-1/4"
4	1"	1"	1-1/4"	1-1/4"	1-1/2"
2	1"	1-1/4"	1-1/4"	1-1/2"	2"
1/0	1-1/4"	1-1/2"	2"	2"	2-1/2"
2/0	1-1/2"	1-1/2"	2"	2"	2-1/2"
4/0	2"	2"	2-1/2"	2-1/2"	3"

ELECTRICAL

As mentioned earlier, if a remodeling job adds a significant number of new circuits or appliances with heavy power demands, the present service is probably not adequate, and the service should be upgraded. This chapter has guidelines for estimating changeover of the service entrance and total power requirements.

Removing Existing Equipment. Three parts of the system should be considered in the estimate: the service entrance, permanently mounted fixtures and present wiring and fittings. Updating the service entrance always involves a direct replacement of existing equipment. Ideally, this should be done as a single operation to avoid lengthy interruption of power. In some cases the present box can be converted to a subpanel rather than removed. See "Service Entrance Installation" in this chapter for estimating guidelines. The labor to remove existing fixtures can vary widely, depending on room location and type of fixture.

Removal	**Minutes Each**
Ceiling/wall lamps and fixtures	15
Base plugs/receptacles (120 V.)	10
Appliance receptacles (240 V.)	15
Switches/dimmer controls	10

Removing existing wiring and fittings can be a complex operation, especially if the remodeling work is a long way from the service entrance, or if the wiring is inaccessible. It may be that existing wires need not be removed, depending on whether they will interfere with new work and the local code requirements.

If the present wiring is metallic or non-metallic flexible cable, it can probably be cut off after the fixture is removed, and no further time need be spent in removing it. If rigid conduit was used previously, its condition should be checked. When there are obvious leaks or fire damage, it should not be used for new wiring. If it is in good condition, and placed appropriately for the remodeling work, it can often be used for new wiring. Even the existing wiring can be used, if it is of adequate size. Where conduit is large enough, new wires can be pulled, and old ones cut off; if this isn't possible, then the old wiring will have to be removed before new wires are pulled. Unless rigid conduit is in the way, it can be left in, even if it will not be reused. Figure the following to remove wire from conduit:

	Less than 25' from jct. box	More than 25' from jct. box
Open wall, ceiling or floor adjacent	15 min. per box (switch, plug, etc.)	30 min. per box (switch, plug, etc.)
No opening area adjacent	30 min. per box	45 min. per box

Careful planning for this part of the job can save time and work later on. Consider whether portions of the present wiring can be reused. It is important that electrical service be brought up to date in accordance with local codes.

Planning Guides
Determining Total Power Needs

When a remodeling project is planned, the total electrical load must be calculated in order to determine whether the service is adequate. The National Electrical Code suggests the following method:

1. Using outside dimensions, multiply the number of square feet of living space by 3 watts.

2. Add 1500 watts for each 20-amp appliance circuit.

3. Add the actual requirement for each major appliance.

4. Subtract 10,000 watts from the total, and multiply the difference by 40%; add the 10,000 watts for a subtotal.

5. Add the actual requirement for air conditioning or heating, whichever is larger.

6. Convert load to current by dividing the final total by 240 volts.

Note: The service entrance panel must have a rating higher than the number of amps calculated.

ELECTRICAL

For a service rating less than 100 amps, proceed as follows:

1. Follow steps 1 and 2 above.

2. To the first 3000 watts, add 35% of the amount over 3000 watts.

3. To the subtotal, add the actual requirements for all major appliances.

4. Convert to current by dividing the total by 120 volts or 240 volts, depending on existing service.

For 100-amp service or higher, you can construct a simple chart:

```
_____ sq. ft. living space x 3 watts      _____ watts
_____ 20-amp circuits x 1500 watts         _____ watts
Laundry circuit (1500 watts)                    _____ watts
Major appliances:    water heater               _____ watts
                     clothes dryer              _____ watts
                     dishwasher                 _____ watts
                     disposal                   _____ watts
                     range                      _____ watts
                     other                      _____ watts
                            Total               _____ watts
                                                - 10,000 watts
                       difference               _____
                           x 0.40               _____
                                                + 10,000 watts
                         Subtotal               _____ watts
                Air cond or heater              _____ watts
                       Total Load               _____ watts
                       ÷ 240 volts              _____ amps
```

If the present service is less than the total number of amps when new circuits are included, the service entrance will have to be updated to accommodate the increased load.

Ampacity of Insulated Copper Conductors

Wire Size	Insulation Type	Ampacity
14	TW, THW, THWN	15
12	TW, THW, THWN	20
10	TW, THW, THWN	30
8	TW	40
8	THW, THWN	45
6	TW	55
6	THW, THWN	65
4	TW	70
4	THW*, THWN*	85
2	TW	95
2	THW*, THWN	115
1	THW, THWN	130
2/0	THW*, THWN	175

*Exception - when used as service entrance conductor:

4	THW, THWN	100
2	THW, THWN	125
1	THW, THWN	150
2/0	THW, THWN	200

Number of Conductors Per Box

Type of Box	Size	Number of Conductors*			
		#14	#12	#10	#8
Octagonal	4" x 1-1/4"	6	5	5	4
	4" x 1-1/2"	7	6	6	5
	4" x 2-1/8"	10	9	8	7
Square	4" x 1-1/4"	9	8	7	6
	4" x 1-1/2"	10	9	8	7
	4" x 2-1/8"	15	13	12	10
	4-11/16" x 1-1/4"	12	11	10	8
	4-11/16" x 1-1/2"	14	13	11	9
Switch	3" x 2" x 2-1/4"	5	4	4	3
	3" x 2" x 2-1/2"	6	5	5	4
	3" x 2" x 2-3/4"	7	6	5	4

3" x 2" x 3-1/2"	9	8	7	6

*Count all grounding wires in a box, each device and each wire entering and leaving the box without splice as one conductor. Pigtails are not counted at all.

Basic Installations. If the service entrance panel is to be replaced, calculating the total power requirements will determine the size equipment needed. Remember, include the cost of the service riser (pipe) or underground conduit riser in the total cost of equipment. Check with the utility company to determine where wires will be brought to the riser. Installation starts there. For underground service it may be necessary to provide a trench and conduit to the utility pole or property line. The utility may bring the trench to a junction box (pull box) a specified distance from the house. Estimate according to the distance you'll have to take the pipe and cable.

To estimate the remaining electrical work, draw up a wiring plan. Determine where circuits are needed, how long each run must be and the size and type of conduit and fittings. It will help in deciding whether the circuits should be wired directly to the service entrance panel, or to a subpanel served by a large branch line. It is also possible to convert the existing service entrance to a subpanel, if the existing circuits will remain in place and the service entrance must be upgraded.

If the number of circuits must be increased, as is often the case, the estimate must include the necessary circuit breaker, conduit and cable or shielded cable and the fittings to be used in the circuit. Calculate the length of circuits and the number of each type of fitting, then the size and length of wire or cable to determine the total material cost. Be sure to add 2' of cable for each box and 6' for connection to a panel or subpanel.

Labor Per Circuit With One Connection
(Average 20')

	Open Wall	**Closed Wall**
Rigid Conduit	$22.27	$37.10
Romex or BX	21.74	29.65

Also important in estimating electrical work is whether the walls are open or closed. In new construction or rooms where wall coverings have not been installed, wiring can be done more quickly. This permits use of rigid conduit, generally considered the safest method. If local code permits, metallic cable is equally safe when properly installed. Non-metallic cable can be installed more quickly, but it is often prohibited by the code.

If walls and ceilings are closed, installation becomes more complicated. Metallic cable is the recommended method, but check local codes. Wiring plans should take into consideration the direction of the joists, and whether the cable can run along the joists or must run through them.

Labor Per Circuit Through Studs
(Average 20')

	Open Wall	Closed Wall
Rigid Conduit	$32.15	$61.80
Romex or BX	29.65	52.65

Boxes and necessary fittings must also be included in the estimate of material. The wiring plan will help determine the number and types of items needed. See the checklist for typical costs. Each new circuit will involve:

1. Running conduit if walls are open and code requires it, or cutting holes for boxes if not, with additional openings at corners or behind baseboard if necessary to install wiring;

2. Pulling wires through conduit or running required cable to boxes;

3. Installing and wiring switches, receptacles or other hardware at each box;

4. Connecting wires or cable at panel or junction box.

Use the following chart to estimate labor:

ELECTRICAL

Estimating Table for Basic Circuit Installation
(Minutes/Dollars)

	Indoors		Outdoors	
	open wall	closed wall	frame wall	masonry wall
For each 10' - 115V				
Rigid conduit & wire:	15	45	15	15*
	$5.60	$16.80	$5.60	$5.60
Shielded cable with ground:	15	30	--	--
	$7.50	$11.25		
For each 10' - 230V				
Rigid conduit & wire	20	50	20	20*
	$7.50	$18.75	$7.50	$7.50
Shielded cable with ground	15	30	--	--
	$5.60	$11.25		

*Add 15 minutes for masonry anchors.

When wiring must go through studs, estimate as follows for each 10':

	open wall	closed wall
Rigid conduit	20	50
	$7.50	$18.75
Shielded cable	15	45
	$5.60	$16.80
	115V	**230V**
For connection at each junction box or outlet box, add:	15**	20
	$5.60	$7.50

For connecting each circuit at
service entrance or subpanel, add .... <u>10</u>   <u>15</u>
                                                                $3.75   5.60

** Add 5 minutes for 3-way switch.

After estimating all circuits, add all installation time and multiply total time by the hourly rate:

	Hours	Rate	Total	Rate	Total
Electrician	10.0	--	--	$29.65	$296.50

When existing conduit will be left in place and wiring upgraded, figure 5 min. for each 10' of hot and neutral wire to be pulled, and an additional 5 min. for each box in the circuit.

When cable is installed across open joists in basements and attics, it may be necessary to provide wooden raceways—local code will specify what is required. Figure 10 min. for each 10' of raceway installed.

Where wiring must be installed in existing walls, surface raceway is also a possibility. It is not recommended for residential use, because it is generally considered unattractive. However, it can be used in a garage or utility room, or other areas where appearance is not a major concern. See the checklist for typical material costs. Figure installation at 15 min. for each 10' of surface raceway, plus 15 min. for installation of each box.

Service Entrance Installation. When added electrical power is needed, include the service entrance material shown on the checklist and the electrician's labor to install it. Whether the wiring is overhead or underground from the utility lines, allow approximately one full day (8 hours) for the changeover.

To install a new service entrance, include the following. Disconnect service entrance cables from utility lines. Disconnect house circuits from panel. Remove conduit or cable connectors from one box and pull wires from box. Pull existing service entrance cable from mast or underground riser. Remove existing mast or riser and install replacement. (If the existing mast or riser can be left in place, deduct two hours from time required to install given below.) Remove service entrance panel and install replacement. Pull new cable through mast or riser. Connect cable to meter socket or main disconnect. Reconnect existing house circuits.

ELECTRICAL

It will be necessary to coordinate the work with the local utility, because their personnel must cut power to the cable before it is disconnected. Likewise, the panel must be inspected before service is reconnected, which will usually involve additional time. Another variable to consider is where the utility will bring the wires, especially for underground installation. Length of the trench, conduit and cable must be included in the estimate. Figure one electrician can complete the preceding in 8 hrs.

	Hours	Rate	Total	Rate	Total
Electrician	8.0	--	--	$29.65	$237.20

If the service from the utility company is adequate to support additional circuits, but the existing service entrance panel cannot accommodate the circuits, it may be necessary to install only a new panel. This may also be the case if the existing panel is damaged or outdated. If only a panel is needed, the following work must be estimated: Disconnect service entrance cables from the main circuit breaker. Disconnect house circuits from panel. Remove conduit or cable connectors from box, and pull wires from box. Remove panel and install replacement. Connect cable to meter socket or main disconnect. Reconnect existing house circuits.

Again, power from the utility lines must be shut off before the new panel can be connected and time must be allowed for the panel to be inspected. Allow one electrician four hours.

	Hours	Rate	Total	Rate	Total
Electrician	4.0	--	--	$29.65	$118.60

When only a few additional circuits are needed, or a single appliance circuit is added, both the power service and the present panel might be adequate. It is often necessary to add circuit breakers to the panel to accommodate the additional wiring. In most panels of recent manufacture, it is relatively simple to add circuit breakers and wire the new circuits in. Allow 15 minutes for adding a circuit breaker to the existing panel.

Another possibility is that the power and the service entrance panel are adequate, but there is no space remaining in the panel for additional circuit breakers. In this case a subpanel is added, either next to the main panel or near the appliance the subpanel will serve. Branch circuits can be run from the subpanel as needed.

The estimate includes the following work items: Install subpanel box. Connect conduit and pull cable, or run heavy cable as provided in local code. Connect cable at subpanel box. Turn off power and connect cable at service entrance panel.

	Hours	Rate	Total	Rate	Total
Electrician	2.0	--	--	$29.65	$59.30

Another practical solution is to convert the existing panel to a subpanel. This leaves circuits intact that are already connected. It can often delay turning off power until most of the rewiring is done and the service is ready to be moved to the new panel. Figure the installation of the new service entrance panel as before.

The conversion of existing panel to subpanel is as follows. Disconnect original service entrance cable and remove wires or fold and tape them out of the way. Remove grounding conductor and bonding screw. Remove branch circuit grounding wires (if any) and connect them to new panel. Run conduit or subfeed cable to new panel. Connect subfeed cable at existing panel and at new service entrance panel.

	Hours	Rate	Total	Rate	Total
Electrician	3.0	--	--	$29.65	$88.95

A final word about installing a new service entrance. This is the best time to change the location of power lines if it is desirable to do so. Be sure to consider whether the present entrance is the best place, or whether new construction or the design of the home dictate a new location for the service entrance. It is best to make all changes at the same time. Careful planning of the job will help to determine whether a change of location should be recommended.

Distribution. After you have determined the service entrance installation, the next consideration is how the power will be distributed within the home. Local codes and the particulars of the job usually determine the type of conductor. For example, in a room addition, where the walls are open, IMC or EMT conduit is sometimes required. Non-metallic cable is the simplest to work with, but codes vary widely on where it can be used. Common sense dictates that it should not be used where the wiring is exposed, especially if there is any danger of damage to the wiring or of people coming in contact with it. Be sure to calculate the cost of conduit and wires when IMC or EMT are used.

Planning the wiring is the next step. Basically, there are three ways to go: through the basement, the attic or existing walls. Labor costs will depend on

ELECTRICAL

which of these routes is available. Since it's a fairly simple job to run cable along joists or through holes in them in a basement or attic, less time will be needed than for working in areas where access is limited.

If wiring needs to run behind a finished wall, check to see whether the wiring can run between the studs. If it needs to be run through studs, extra time is necessary for opening the wall and making holes through the studs. A safe and practical shortcut to look for is back-to-back boxes. If a new box can be connected into another where the wiring is already in place, much less time is required for the installation. Remember, too, that if strips and other access openings are made in the wall or ceiling, it's not usually the electrician's job to repair them. See the chapter on installing and repairing wall coverings.

Wiring size depends on the type of circuit. Later in this chapter, there are recommended amperage ratings for circuits most likely found in the home. The type of wire will also vary depending on code requirements, and on whether it is a dedicated circuit for an appliance or other heavy-duty use.

Outdoor wiring presents some special considerations in estimating. When installing an air conditioner, hot tub or other outdoor appliance, it is often most practical to run the circuit from the service entrance along the outside of the house. This installation generally requires rigid conduit or IMC, and a small subpanel with circuit breakers. Check local code requirements and be sure to include any special equipment needed when preparing the estimate.

TYPICAL ELECTRICAL WORK IN REMODELING

This section provides recommendations for planning and estimating the kind of electrical work that is typical in residential remodeling.

Living Room, Dining Room and Hallways. Unless there are special needs that must be figured separately, work in these rooms will usually involve simple 115-volt circuits for ceiling and wall lighting fixtures, base plugs or other receptacles and switches. A 15-amp circuit with grounded receptacles is recommended, with an additional 20-amp circuit for the dining room to allow for use of appliances requiring fairly large amounts of current. Light fixtures should not be connected to the 20-amp circuit.

In the living rooms or hallways, at least one receptacle should be wired to a switch for convenience in turning a lamp on or off from the entrance. This can be a split receptacle, with only one of the plugs controlled from the switch.

These circuits and all those that follow can be figured as shown in the *Basic Installations* section of this chapter. Add 15 minutes for each split receptacle. Use the following tables to estimate simple circuits. For wiring only on an average hallway, including ceiling fixture, switch and receptacle:

	Hours	Rate	Total	Rate	Total
Open Wall	2.5	--	--	$29.65	$ 74.13
Closed Wall	4	--	--	29.65	118.60

Add 0.5 hour for each 10' of circuit that runs through joists.

For wiring only in the average living or dining room, including one switch and four receptacles (one split) in the living room, or one switch, one ceiling fixture, two receptacles and one 20-amp receptacle in a dining room:

	Hours	Rate	Total	Rate	Total
Open Wall	4.5	--	--	$29.65	$133.43
Closed Wall	7.5	--	--	29.65	222.38

Add 0.5 hour for each 10' of circuit that runs through joists.

Bedrooms. Ceiling and wall lighting fixtures, receptacles and switches are usually all that is needed in a bedroom, and 115-volt, 15-amp grounded circuits are recommended. If extensive wiring is planned for the master bedroom, a central control unit can be installed that switches lamps and appliances on and off in other parts of the house.

For convenience at least one outlet per bedroom should be controlled by a switch near the door. Special consideration should also be given to special requirements, such as closet lights with switches outside the closet and lighting or convenience outlets near beds.

For wiring only, which includes switch and four receptacles, in an average bedroom, an electrician's labor will cost:

	Hours	Rate	Total	Rate	Total
Open Wall	4.0	--	--	$29.65	$118.60
Closed Wall	7.0	--	--	29.65	207.55

Add 0.5 hr. for each 10' of circuit that runs through joists.

Bathrooms. Circuits of 115-volt, 15-amp are recommended for ceiling and wall lighting fixtures and convenience outlets. Multiple boxes should be planned for dramatic lighting over tubs or vanities and for exhaust fans, sunlamps and auxiliary lighting over counters or in powder room areas.

When it comes to the bath, a customer might want the works—whirlpool, sauna or both. It will be necessary to include a 230-volt, 40-50 amp circuit for heaters. The plan must include the necessary switches for whatever equipment is installed, as well as the placement of the switches. Be sure to calculate total current needs carefully so that adequate circuits can be provided. Cost of individual fixtures will vary depending on the unit. Check with the supplier before preparing the estimate.

For wiring only, including ceiling or wall fixture, switch and receptacle, in the average bathroom, labor cost is as follows:

	Hours	Rate	Total	Rate	Total
Open Wall	2.25	--	--	$29.65	$ 66.71
Closed Wall	4.25	--	--	29.65	126.01

For a 230-volt circuit, add:

	Hours	Rate	Total	Rate	Total
Open Wall	1.25	--	--	$29.65	$37.06
Closed Wall	2.5	--	--	29.65	140.63

Add 0.5 hr. for each 10' of circuit that runs through joists.

Kitchens. The kitchen often presents the greatest challenges in electrical work and requires careful planning. Both for reasons of safety and because it is often the focal point of family living and entertaining, the kitchen needs special attention. Circuits designed for the kitchen should all be 115- volt, 20- amp or 230-volt, 20-50 amp, depending on the appliance. A 15-amp circuit can be included but for lamps only.

Planning is so important for the kitchen, because it presents a number of specialized requirements that must be taken into account when estimating. The basics are there, of course: ceiling and wall lighting fixtures, base plugs and receptacles for countertop use and the usual switches. A range hood or ventilating fan has become standard, and if the range is electric, then a separate high-amperage circuit must be provided for it.

Likewise, 20-amp circuits should be provided for the dishwasher and for the garbage disposal. Check code requirements for placement of the switch controlling the disposal. Usually, it must be a minimum distance from the appliance for safety. If the refrigerator/freezer is a large unit, it may even be advisable to provide a separate circuit for it.

While the kitchen is in the planning stages, consider also that other popular items may be added, such as a trash compactor or a food processor built into the countertop. And it's not unusual for a microwave oven to need 10 amps or more, requiring a circuit that is adequate for the load.

All of these needs can and should be provided for, when figuring the electrical work in the kitchen. Figure each circuit according to the guidelines in the *Basic Installations* section. Also, see the checklist for any special fitting and receptacle needs.

For wiring only, including ceiling fixture, switch and three receptacles, in the average kitchen, the labor cost is as follows:

	Hours	Rate	Total	Rate	Total
Open Wall	2.5	--	--	$29.65	$ 74.13
Closed Wall	4	--	--	29.65	118.60

For stove or other 230-volt circuit, add:

	Hours	Rate	Total	Rate	Total
Open Wall	1.25	--	--	$29.65	$37.06
Closed Wall	2.5	--	--	29.65	74.13

Add 0.5 hr. for each 10' of circuit that runs through joists.

Utility Room. Because of the appliances found here, this room is likely to need circuits somewhat out of proportion to the size of the room. Wall outlets, for instance, should be 115-volt, 20-amp grounded circuits to accommodate a clothes washer and gas dryer.

Separate 230-volt circuits are needed for the clothes dryer and the hot water heater, if they are electric. Usually the dryer will require 20 to 30 amps, and the heater will need 50 amps. A circuit for central air conditioning is another frequent addition to utility room service. It will also require 230 volts and 40 amps, and most codes require a circuit breaker in a box with a cut-off switch within a few feet of the compressor unit.

Be sure to allow for a switchbox connection adjacent to the furnace. Although it doesn't usually require high amperage, the furnace circuit and switch are often spelled out very specifically in electrical codes. Allow an additional 15 minutes for the air conditioning circuit if it must go through a wooden frame wall, or an additional 30 minutes if the wall is cement block or masonry.

ELECTRICAL

For wiring only, in the average utility room, including ceiling fixture, switch and two receptacles, the cost is as follows:

	Hours	Rate	Total	Rate	Total
Open Wall	3.0	--	--	$29.65	$88.95
Closed Wall	5.0	--	--	29.65	148.25

For each 230-volt circuit, add:

	Hours	Rate	Total	Rate	Total
Open Wall	1.0	--	--	$29.65	$29.65
Closed Wall	2.0	--	--	29.65	59.30

Add 0.5 hour for each 10' of circuit that runs through joists.

Family Room or Den. Circuits will be needed for ceiling fixtures—especially if the room is on a previously unfinished lower level—and wall lighting, wall outlets and switches. Receptacles may be required for special areas such as a fireplace, a bar or in an area where an electronic game or jukebox will be placed.

For the most part, circuits to this type of room can be 115-volt, 15-amp grounded. If a refrigerator or other heavy appliances are a possibility, then a 20-amp circuit should be included. At least one wall receptacle controlled by a switch is also a good idea.

For wiring only, in an average family room, including ceiling fixture, switch and four receptacles, the cost is as follows:

	Hours	Rate	Total	Rate	Total
Open Wall	4.5	--	--	$29.65	$133.43
Closed Wall	7.5	--	--	29.65	222.38

Add 0.5 hour for each 10' of circuit that runs through joists.

For stereo or TV wiring, figure the labor as comparable to installing a standard 115-volt circuit. Consult the manufacturer or retailer for the proper wiring and receptacles to be used.

Garage and Outdoors. Adding or updating circuits in the garage will often be a simpler task than elsewhere for two reasons. It is often the location, or near the location, of the service entrance, and walls are often open, simplifying the wiring job. Circuits will usually be needed for ceiling fixtures, especially in a workshop area. A 115-volt, 15-amp service is adequate. Three-way switches are desirable so that lights can be controlled from the door into the house as well as near the garage door itself.

Other receptacles in the garage should be 115-volt, 20-amp circuits to allow for use of tools and appliances. It may also be necessary to include a 230-volt circuit with a special plug to meet the customer's particular needs. When the wiring will be left exposed, it is wise to specify IMC or metallic cable, even if the local code does not require it.

Frequently, special wiring needs are present in a garage, such as for an electric door opener or even a doorbell transformer. These can usually be connected to existing boxes for light fixtures, but must be included in the estimate if additional work will be necessary.

Outdoor receptacles should also be part of the plans. If fixtures are to be mounted near the garage doors, or floodlights on the soffits, circuits will have to be provided. Standard wiring and boxes will usually be adequate, since the fixtures will protect them from weather. Weatherproof receptacles should also be provided for convenience outside the garage, on a patio or elsewhere as needed. See the checklist for suggested special hardware that will be required for them.

Figure the work for garage and outdoor circuits according to the guidelines in the *Basic Installations* section of this chapter. If an outdoor receptacle is mounted on a frame wall, add 15 minutes for each box; on a masonry wall, add 30 minutes for each box. The following tables can be used for simple circuits. For wiring only, for an average garage, including ceiling fixtures, switch and two receptacles, it should cost:

	Hours	Rate	Total	Rate	Total
Open Wall	3.0	--	--	$29.65	$ 88.95
Closed Wall	5.0	--	--	29.65	148.25

For 230-volt appliance circuit, add:

	Hours	Rate	Total	Rate	Total
Open Wall	2.0	--	--	$29.65	$59.30
Closed Wall	3.0	--	--	29.65	88.95

Add 0.5 hour for each 10' of circuit that runs through joists.

Special Projects. There are many wiring jobs that do not relate specifically to one room in the house, or are not ordinarily part of a remodeling project. Even where the customer doesn't request these installations, it may be wise to suggest one or more of them for the improvement of the house.

ELECTRICAL

Ceiling Fans typically cost $60.00 to $300.00. These can be a valuable addition to a home because of the energy savings they provide. In many cases, they can be added to an existing circuit that served a ceiling lamp fixture. If not, a 15-volt, 15-amp grounded circuit with a switch is all that is required, a fairly simple addition if other electrical work is being done in the room.

Intercom or Stereo, costing $200.00 to $400.00, is an installation that does not usually require an electrician, but it can be done along with other electrical work. Since heavy current loads are not involved, the wiring is usually light duty; check the manufacturer's recommendations. If extensive remodeling is being done, or a somewhat isolated room is being added, the stereo or intercom may be a welcome extra. Figure the wiring as similar to installing a 115-volt line not requiring conduit and estimate it accordingly.

Alarms and Remote Controls cost about $50.00 to $200.00. For these projects the work will vary with the complexity of the system. Consult the local electrical code for requirements relating to smoke, fire and burglar alarms. The work will be similar to installing 115-volt lines, and one of those will often be necessary to furnish power to the alarm system. Remote controls may involve wiring into each circuit to be controlled from the master station. If so, figure the length of each run and estimate as you would for other circuits, taking into account the type of wiring needed and the number of connections.

Auxiliary Heating Units typically cost $30.00 to $100.00. These are especially useful in room additions, or in attics and basements. These units will usually require an additional 230-volt, 40-amp isolated circuit. For a gas-fired heater with an electric fan, the circuit rating can be less. A 115-volt, 15-amp circuit is adequate.

Electronic Air Cleaners cost about $175.00 to $250.00. This is worth considering in a home that must remain closed most of the time for heating and air conditioning. Since these are most frequently mounted on the furnace, they can be connected to an existing circuit in the utility room or basement, or to the circuit that serves the furnace itself.

Low-Voltage Lighting usually costs $75.00 to $125.00. It is an excellent idea for accenting decorative planting and lighting walkways. A transformer must be mounted outside the house, and a suitable UF cable must be run to the lights. Check manufacturers' specifications for required wiring, and figure the work as similar to other circuits. Be sure to allow for trenching or burying the cable to the various location of the lamps.

Outdoor Fixtures cost $10.00 to $75.00 and range from spotlights mounted on the house to post lamps, garden lighting or convenience outlets for any purpose. The major difference in estimating outdoor wiring is in figuring the type of materials to be used. Rigid steel or PVC conduit is usually required, and if plastic is used, ground wires must be provided. Driptight or watertight panels, switch-boxes and receptacles will also be needed.

Electrical codes require that new outside receptacles be protected with a GFCI. Wiring these and other outdoor circuits can be figured using the guidelines in *Basic Installations*. The estimate may also have to include time for running underground wiring when necessary. If the circuit will connect to an existing box, add 15 minutes for adding an extender ring. If power will be brought from a box inside the house, add 15 minutes to bring wiring through a frame wall, or 30 minutes for a masonry wall. For each 10' of circuit:

	Hours	Rate	Total	Rate	Total
Open Wall	1.0	--	--	$29.65	$ 29.65
Closed Wall	1.5	--	--	29.65	44.48
Average air-conditioner circuit	5.0	--	--	29.65	148.25

Add 0.5 hour for each 10' of circuit that runs through joists.

Basements and Attics. Electrical installation in these areas will be similar to new construction—the walls are open and existing circuitry is reasonably accessible. Depending on appliances to be used, either 115-volt or 230-volt circuits will be needed. The 15 to 20 amp circuits should be included for general purposes, and heavier circuits are needed for large appliances.

When an attic is remodeled, usually it is converted to a bedroom or other living space. The plan should include circuits for lighting, receptacles and possibly a ventilator. If auxiliary heating is needed, a separate circuit of the proper size must also be figured.

For basement remodeling, 115-volt, 15 or 20 amp circuits will be needed for lighting and receptacles. A 230-volt, 20 to 50 amp circuit will be needed if the utility room is part of the project. Kitchen and bathroom installations are frequently part of a basement project. Whether basement or attic, the circuit plan is essential in estimating the work involved.

For wiring only, in an average basement or attic, including ceiling fixture, switch and four outlets, labor cost is as follows:

	Hours	Rate	Total	Rate	Total
Basement	4.0	--	--	$29.65	$118.60
Attic	7.0	--	--	29.65	207.55

For 230-volt air-conditioning or appliance circuit, add:

ELECTRICAL

	Hours	Rate	Total	Rate	Total
Basement	2.0	--	--	$29.65	$ 59.30
Attic	3.5	--	--	29.65	103.78

Add 0.5 hour for each 10' of circuit that runs through joists

Grounding and GFCIs. Grounded circuits are highly recommended. In most instances electrical codes require that circuits be grounded and that grounded outlets be provided. Even when they are not required, they are advisable for the safety of your clients.

Protection afforded by a ground fault circuit interrupter (GFCI) is recommended and required for many outdoor installations. They are a good idea for any circuit where a hand-held appliance will be used. For the small amount that they increase the cost of the job, they are valuable protection. Typical costs run $35.00 to $50.00.

Electrical Estimate Checklist

Service Entrance   Cost
(200 Amp Service 2" Pipe)
Service Entrance Head Mast (Service Riser) .... $ _____ ea.
Underground Meter Socket .... _____ ea.
Roof Tie .... _____ ea.
Hub .... _____ ea.
2-Hole Pipe Straps .... _____ ea.
1/2" Ground Clamp (42 CKT 200 Amp Main Breaker) .... _____ ea.
Service Entrance Panel .... _____ ea.
              2-Pole .... _____ ea.
Circuit Breakers .... _____ ea.
Wire Connectors 73B .... _____ ea.
Locknuts .... _____ ea.
Bushings .... _____ ea.

Single Conductors
Low Voltage #18 MTW .... _____ /m
Standard 12 TW .... _____ /m
Heavy Duty (3/0) .... _____ /m

Multi-conductor Cable
14-2 Type NM (Non-Metallic Sheath) .... _____ /m
14-2 Type UF (Underground Feeder) .... _____ /m
8/3 Type SE (Service Entrance) .... _____ /m

Conduit 1/2"
1/2" Intermediate Metal (IMC) .... _____ /c
1/2" Couplings (Threaded) .... _____ /c
1/2" No Thread Compression Connectors .... _____ /c
Thinwall Metal (EMT) .... _____ /c
1/2" Couplings (Set Screw) .... _____ /c
1/2" Couplings (Compression) .... _____ /c
1/2" Connectors (Set Screw) .... _____ /c
1/2" Connectors (Compression) .... _____ /c
3/8" Flexible Metal Conduit .... _____ /c

ELECTRICAL

(BX) Connector - Angle Flex .... _____/c
Takeall Connector Adaptors .... _____/c

Boxes and Accessories
(4 x 4) Extender Ring.. _____ ea.
(4 x 4) Switch Box w/ Bracket .... _____ ea.
(2 x 4) Gem Box - Removable Sides .... _____ ea.
(2 x 4) Switch Cover.. _____ ea.
(2 x 4) Duplex Cover .... _____ ea.
Outlet Box:
    Octagonal .... _____ ea.
        Plain Cover .... _____ ea.
    Square Box .... _____ ea.
        Plaster Rings .... _____ ea.
        Raised Covers .... _____ ea.
Junction Box:
    Octagonal/Cover... _____ ea.
    Square/Cover.. _____ ea.
Special Boxes:
    Weatherproof .... _____ ea.
    Switch Cover .... _____ ea.
    Receptacle Cover .... _____ ea.
    "Handy" Box .... _____ ea.
    Covers .... _____ ea.
    Gem Box (for drywall) .... _____ ea.

Switches
Single Pole.. _____ ea.
Three-way.. _____ ea.
Four-way .... _____ ea.
Dimmer Incandescent .... _____ ea.

Receptacles
Duplex - Grounded.. _____ ea.
Recessed (Clock)... _____ ea.
Special:
    Air Conditioner or Clothes Dryer .... _____ ea.
Kitchen Range... _____ ea.

Circuit Breakers
Slide-in .... _____ ea.
Receptacle (GFCI) .... _____ ea.

Miscellaneous
Conduit Straps .... _____ ea.
Staples .... _____ ea.
Wire Connectors (T&B) .... _____ ea.

Chapter 11

HVAC

There have been many innovations in heating and cooling technology in recent years, especially in residential work, due to the research programs conducted by heating and air conditioning associations, engineering societies, equipment manufacturers and universities.

Conventional Heating Systems. There are four basic types of conventional heating systems—warm air, hot water, steam and electric, including both resistance types and heat pumps, each of which is used separately or in combination with the others.

For small to medium residences, stores, churches and the like, warm air systems are most common. For medium to large residential and non-residential buildings, hot water systems are often selected. For large-scale installations steam systems are common. Inasmuch as there are no definite rules stating just when to select one system in preference to another, there is much overlapping in their application. An owner's preference for a particular system might be governed by the initial cost of installation, operation cost, convenience of operation, adaptability of system to other uses, such as summer cooling using ducts of a warm air system, snow melting using hot water piping connected to a hot water system, fuel availability and aesthetics.

Panel Heating Systems. Panel heating systems are used in residential work and for some non-residential structures as well. These systems use large areas of room surfaces, heated to relatively low temperatures (80° to 100°F.), as radiant panels. Panels usually are heated by warm water piping, warm air ducts or low temperature electrical resistance elements embedded in or located behind ceiling, wall or floor surfaces.

The warm water piping method uses piping embedded in concrete floors or plastered ceilings. In some cases, both floors and ceilings are used as radiant panels. Piping in floors is usually welded joint black wrought iron pipe, steel pipe or copper tubing. For installations in plastered ceilings, copper tubing is usually used.

Estimating the Cost of Heating Systems. There is no fast, easy rule of thumb for estimating the cost of a heating system. This requires expert knowledge, because most jobs must be designed and laid out before they can be estimated. In residential and small commercial work, architects seldom give more information than merely specifying the type of system desired, the performance expected and the kinds of material to use. It is then up to the heating contractor to figure the heating loads and design a system to fit the conditions. On larger work the heating system is usually designed and laid out diagrammatically, but the heating contractor must work out the details, such as for piping equipment connections and controls.

The necessary technical knowledge for this purpose is acquired through everyday participation in this line of work and constant study to keep up to date. Heating, ventilation and air conditioning is too complex a subject for detailed treatment in a volume of this kind. Much useful data and information can be gained from manuals published by the professional associations. The information below is for "roughing" or approximate estimates only.

Boilers are designed to burn specific fuel—hard coal, soft coal, oil or gas. Cost varies depending on boiler size and accessories, such as domestic hot water heaters or thermostatic controls.

The contractor encounters cast iron radiators in older residential buildings, usually apartment buildings. What is more common nowadays is the concealed convection type radiators or baseboard radiation.

Copper piping is used for both heating and plumbing. The cost of this work depends to a large extent on the experience of the contractor. Mechanics familiar with sweating joints and installing copper piping can install this work rapidly, while inexperienced workers will need considerably more time. The estimator must be very familiar with the requirements of the trade in order to prepare intelligent estimates.

Determining Heating Loads. The first step in designing and estimating heating work is to determine the maximum heating load. For most residential work the maximum heating load is the total heat loss of the structure figured at design temperatures.

Heating loads and heat losses are expressed in Btu per hour. A Btu (British thermal unit) is the quantity of heat required to raise the temperature of 1 lb. of water 1°F. Building heat loss varies considerably depending on climate, design temperatures, exposure, building size, architectural design, purpose of structure, construction materials and quality of construction workmanship, to name but a few of the factors.

Heat losses should be figured separately for each room so that the proper amount of radiation can be provided for comfort in all areas. Complete information and data on this subject are beyond the scope of this text, and the contractor should consult association manuals and guides for this type of information.

HVAC

FORCED WARM AIR HEATING SYSTEMS

Forced air is the most widely used residential heating method. Generally speaking, forced air differs from the old gravity type by using blower equipment to circulate the air in the system. It is practically impossible for the average general estimator to compute the cost of a modern forced warm air heating system, unless he has a working knowledge of this subject and is able to calculate building heat losses and design systems that will satisfy heating requirements under design conditions. Always obtain firm quotations from reputable heating contractors before submitting bids on jobs containing this work.

Forced warm air heating systems that are equipped with automatic humidifying devices and air filters are popularly known as *winter air conditioning* systems. Most forced air installations are of this type. Each system consists of a direct-fired heating unit, a blower, a system of warm air ducts and a return air duct system. For the heating unit, heat pumps are the preferred choice. They are the most efficient. Air filters are located in the return air duct system just ahead of the blower. Humidifying equipment is placed in the warm air plenum chamber over the furnace unit. Air-cooling equipment can be added to make this a year round system.

The unit can be gas, oil or coal fired. Gas or oil units that are designed for the particular fuel operate at high efficiency, though satisfactory if less efficient results are obtained by using conversion burners installed in units primarily designed for solid fuel.

An automatic thermostat controls the supply of heat and should be located in the living portion of the home. Actuated by limit controls, the blower commences operation automatically when the heated air in the generating unit has reached a pre-determined temperature and cuts out after the source of heat has been shut down and the temperature in the bonnet dropped to a pre-determined point.

Design of air conveying systems of this kind presents an engineering problem that can only be solved with a reasonable degree of accuracy by a competent designer. The heat loss of the building, room by room, is established and the volume of heated and conditioned air required for indoor comfort establishes data from which a duct system can be designed with due regard for velocities, static pressures and delivery temperatures.

Items to be Included in the Estimate. Costs vary considerably and methods of estimating differ, but every estimate for a winter air conditioning system should include the following items:

1. Cartage of material and equipment from shop to job.
2. Winter air conditioning unit, complete with humidifier, air filters and controls.
3. Labor assembling and setting up unit.
4. Oil storage tank installation, including fill and vent piping, oil booster pump if required, oil gauge, piping to unit and oil filter, for oil-fired units, or piping for gas-fired units.
5. Installation of smoke pipe, fittings and accessories.
6. Warm air and return air plenum chambers.
7. Warm air and return air duct systems.
8. Diffusers, registers, intakes and grilles.
9. Special insulation.
10. Electrical work.
11. Labor starting plant in operation and balancing system.
12. Service allowance.
13. Miscellaneous costs.
14. Overhead and profit.

Cartage. The cost of trucking material and equipment from shop to job varies depending on size of job and distance from shop. An allowance of $60.00 to $120.00 should cover this item on most small jobs.

Cost of Winter Air Conditioning Unit. Sizes, capacities and prices of units vary with different manufacturers, but the following listings are representative of models available for residential use.

Approximate Net Prices of Winter Air Conditioning Units Complete with Automatic Controls, Humidifier and Air Filters

Bonnet Rating Btu per Hr.	Register Rating Btu per Hr. Gas Fired Units	Net Price Heating Only	With Cooling	
60,000	48,000	$500.00	24 MBH	$1,870.00
76,000	60,800	530.00	29 MBH	2,295.00
90,000	72,000	560.00	36 MBH	2,465.00
108,000	86,400	640.00	47 MBH	2,975.00
120,000	96,000	820.00	47 MBH	3,145.00
150,000	120,000	1,060.00	58 MBH	3,910.00
180,000	144,000	1,190.00	--	--

Bonnet Rating Btu per Hr.	Register Rating Btu per Hr. Oil Fired Units	Net Price Heating Only		With Cooling
84,000	71,400	890.00	36 MBH	2,465.00
112,000	95,200	1,020.00	42 MBH	2,975.00
140,000	119,000	1,200.00	47 MBH	3,400.00
175,000	148,750	1,360.00	--	--
210,000	178,500	1,530.00	--	--

Assembling and Setting Up Unit. Most gas fired units can be assembled and set up by a sheet metal worker and a helper in 3 to 4 hours. For an oil-fired unit, figure 4 to 5 hours for this work.

Oil Storage Tanks. Common practice for oil storage tanks is the use of one or two inside storage tanks with a capacity of 275 gals. each. Occasionally outside buried tanks of 550 or 1,100 gallon capacity are used. On a contract basis these tanks installed, including all piping, cost as follows:

1-275 gallon inside tank	$ 400.00
2-275 gallon inside tank	780.00
1-550 gallon outside tank	600.00
1-1,000 gallon outside tank	1,000.00

In the use of an outside tank, a basement wall-type pump costing from $60.00 to $80.00 must be used with some types of burner equipment.

Gas Piping. For average installations an allowance of $200.00 to $300.00 should cover cost of material and labor for connecting gas to the unit within the heating room.

Smoke Pipe. Material and labor for smoke pipe connection from unit to flue should cost $80.00 to $120.00 for gas and oil heating units.

Plenum Chambers. Plenum chambers must be fabricated to order in the shop and are usually made from light gauge galvanized steel. They are mounted on top of the furnace and serve to connect the warm air and return air duct systems to the unit. Two plenum chambers are required for each installation. Plenums cost about $8.00 per sq. ft. in 18 ga., $10.00 in 16-ga. and $12.00 in 14 ga.

Warm Air and Return Air Duct Systems. The cost of ductwork varies greatly, depending on the size of job and type of distribution. The warm air heating contractor usually measures and lists each type and size of duct, fitting and accessory. These quantities are priced for material and then an estimate is made of labor required for installation. This is a lengthy operation and requires an experienced heating professional.

When pressed for time, you can approximate the duct system cost on an outlet basis, using a unit price that covers both material and labor. For ordinary installations the following costs per outlet are about average where sheet metal worker wages are from $20.00 to $25.00 per hour:

Type of System	Cost per Outlet
Conventional system—warm air outlets located on inside walls.	
1st floor outlets	$100.00 to 125.00
2nd floor outlets	140.00 to 160.00
Capped stacks	40.00 to 50.00
Radial perimeter system in small 1-sty. buildings with basement or crawl space.	
Low velocity—6" to 8" dia. ducts	$100.00 to 125.00
High velocity—4" to 5" dia. ducts	95.00 to 120.00
Large central return air duct	130.00 to 140.00
Radial perimeter system in small 1-sty. basementless buildings	
Galv. steel sheet metal pipe and fittings	132.00 to 140.00
Fiber duct with galv. steel fittings	132.00 to 140.00
Trunk and branch perimeter system in medium-size, 1-sty. buildings with basement or crawl space.	
Std. galv. steel, rectangular ducts	150.00 to 165.00

Add for items such as diffusers, registers, intakes and special insulation. Excavation, backfill and concrete encasement where required, not included.

Diffusers, Registers and Intakes. Diffusers and registers generally are of pressed steel construction with prime coat finish, are fully adjustable and are available in floor, baseboard or wall types. Return air intakes are of same construction and finish but are non-adjustable.

Floor diffusers cost $16.00 to $21.00 for common sizes. Baseboard diffusers cost $15.00 for 2'-0" lengths and $21.00 for 4'-0" lengths. Wall registers cost $7.00 to $15.00. Wall intake grilles cost $9.00 to $18.00.

Special Insulation. Ducts in unexcavated spaces or in unheated attic spaces, both supply and return, should be insulated with not less than 1/2" of adequate insulating material. This material is furnished in flexible form and averages $2.00 to $2.50 per sq. ft. applied.

Electrical Work. There are many variations on electrical circuits from meter to fused safety switch adjacent to the heating unit. The same is true for the connection of the blower motor to 110-volt controls and for the 22-volt wiring to a thermostat. Most often the job electrician runs the 110-volt service to the safety switch adjacent to the heating unit. From this point it is handled as a subcontract under the heating contractor and represents a cost from $90.00 to $225.00 for complete wiring installation of all controls and motors.

Labor Starting and Balancing Systems. Normally, the heating contractor starts the heating plant in operation, tests all controls and balances the air distribution system. For jobs completed during the summer months, this might mean a comeback call at the beginning of the heating season. Labor costs for this work will be about $100.00 to $150.00 depending on the size of the job.

Service Allowance. Most jobs carry a one-year free service warranty and an allowance should be included it the estimate to cover this contingency. Average costs for service during the first year of operation are $200.00 for gas-fired and $220.00 for oil-fired systems.

Miscellaneous Costs. Under miscellaneous costs are classified such items as federal tax, sales tax and permits for oil burner or for installation if required.

Overhead and Profit. Most heating contractors would like to add 10% for overhead and 10% for profit to their estimates, but in a competitive market the majority of jobs are being quoted with a straight 10% markup.

Approximate Cost of a Complete System Installed. Assuming installation in residences of ordinary construction located in large cities, the cost of average winter air conditioning systems ranges as follows for gas-fired units. For oil-fired units, add 10% to 15%.

Floor Area Sq. Ft.	Heat Loss Btu per Hr.	Price for Complete Gas-Fired Installation
Conventional System, 1-Story Building with Basement or Crawl Space		
1,000	65,000	$2,550 to 3,400
1,500	80,000	3,400 to 3,910
2,000	100,000	3,910 to 4,250
Conventional System, 2-Story Building with Basement		
1,500	75,000	$3,400 to 3,910
2,000	90,000	4,250 to 5,100
Perimeter System, 1-Story Building with Basement or Crawl Space		
1,000	65,000	$2,550 to 3,400
1,500	80,000	3,400 to 3,910
2,000	100,000	4,250 to 4,760

Perimeter System, 1-Story Basementless Building

1,000	65,000	$2,550 to 3,400
1,500	80,000	3,400 to 3,910

The above prices are for individual jobs. For multiple housing projects, where large numbers of similar dwelling units are involved, the contractor will probably obtain much better prices. Equipment, design and practices vary greatly, even within a local area. No safe guide can be set down for general use. So much depends on correct design, proven equipment and experienced installation that it behooves the buyer to carefully investigate all of these items before awarding a contract for this important part of the building.

SUMMER AIR CONDITIONING

Where forced air heating is used, the same air distribution duct system can be used for both heating and cooling. Various warm air heating equipment manufacturers make cooling units for this purpose, and installation work is usually done by heating contractors.

The function of summer air conditioning equipment is the reverse of winter air conditioning—circulating air is cooled and dehumidified—but design problems are similar. However, for cooling they tend to be more complex. A basic requirement for designing a cooling system is calculating the total cooling load or total heat that must be removed from the structure to achieve a predetermined inside temperature and humidity under design conditions. Heat gain calculations are much more involved than those for heat losses, because many additional factors must be considered, such as the sun effect on roof, walls and glass areas; or internal heat gains from human occupancy, cooking, lighting, appliances and latent heat energy involved in dehumidification.

Heat gain calculations are always expressed in Btu per hour. Cooling equipment capacities are usually stated per ton of refrigeration. A ton of refrigeration will remove heat at the rate of 12,000 Btu per hr. This is equal to the heat required to melt a ton of ice in 24 hours. With water-cooled compression refrigeration equipment, one horsepower usually equals about one ton of cooling capacity.

Types of Cooling Equipment for Residential Use. There are several types of cooling units available for residential summer air conditioning. In general, equipment is classified as water-cooled or air-cooled, with each type available for several methods of application.

One method employs a remotely located condensing unit, with refrigerant lines connected to a cooling coil installed in the warm air plenum chamber of a forced warm air heating system. This method uses the heating system blower and ducts for air distribution.

Another method uses a self-contained unit, with the condenser, cooling coil and blower combined in one cabinet. This type is either cut into the heating ducts or has its own duct system where hot water or steam heat is used.

Net Prices of Remote System Cooling Units. Approximate prices are for factory-assembled units with heating and cooling thermostat but with no pipe, duct materials or blower assembly included.

Capacity Tons	Rating Btu per Hr.	Net Price Air-Cooled	Labor to Install
2	24,000	$1,000.00	8 hrs.
3	36,000	1,450.00	14 hrs.
5	60,000	2,700.00	30 hrs.

Where a complete air circulation duct system is required, figure it the same as given for winter air conditioning systems.

Approximate Cost of Complete Cooling Unit Installation. Based on conditions of average difficulty, complete residential cooling unit installations should cost approximately as follows:

Capacity Tons	Rating Btu per Hr.	Price for Complete Installation
	Remote, Water-cooled System	
2	24,000	$3,740 to 4,250
3	36,000	4,250 to 4,760
	Remote, Air-cooled System	
2	24,000	$3,400 to 3,740
3	36,000	4,250 to 4,590
5	60,000	7,650 to 8,500
	Self-contained, Water-cooled System	
2	24,000	$3,910 to 4,250
3	36,000	5,950 to 6,290
	Self-contained, Air-cooled System	
2	24,000	$2,600 to 4,760
3	36,000	7,650 to 8,160
5	60,000	11,900 to 12,750

Room-size Air Conditioning Units. The cost of portable room air conditioners varies according to size, ranging from 1/2-Hp, which cools room areas up to 400 sq. ft., to 2-Hp for spaces up to 1,200 sq. ft. All models are for window installation and include automatic thermostat controls. Net price is from $400.00 for a 4,000-Btu unit to $1,500.00 for 20,000-Btu unit. If electric wiring is necessary, add extra for installation.

STEAM AND HOT WATER HEATING

Because there are so many conditions and factors that affect the installation of heating apparatus, it is impossible to come up with a set of rules that covers all buildings. It is necessary to consider all conditions in and around the building and additions and deductions made to suit the requirements.

Methods of Computing Heating. The most advanced method of figuring heating, the one that is generally used by heating engineers, is the *Btu method*. This method is based on the concept of replacing the heat loss through exposed walls, doors and windows, floors and ceilings. So many factors are involved in this computation that it is far beyond the scope of this book.

Approximate Prices for Copper Tubing

The price of metals such as copper fluctuates widely. Always check prices before submitting bids.

	Nominal Size							
Price per lin. ft.	1/4"	3/8"	1/2"	5/8"	3/4"	1"	1-1/4"	1-1/2"
Type "K"	0.56	0.91	1.19	1.40	2.09	2.52	3.15	4.06
Type "L"	0.52	0.70	1.05	1.26	1.54	2.10	2.03	3.50

Price per lin. ft.	2"	2-1/2"	3"	3-1/2"	4"	5"	6"
Type "K"	6.09	8.68	11.90	16.80	21.00	49.00	70.00
Type "L"	5.25	7.56	10.15	13.72	16.80	39.20	53.20

When purchased in quantities of more than 2,000 ft. or lbs., whichever is greater, deduct 12% from the above prices.

Prices of Brass Solder-Joint Valves

These valves are made for use with copper tubing and can be used for steam or water service.

Size in Inches

Description of Fitting	1/2"	3/4"	1"	1-1/4"	1-1/2"	2"
Gate Valve	15.61	19.95	24.15	32.20	41.65	58.50

Approximate Quantity of Solder and Flux Required to Make 100 Joints

Size, Inches	3/8"	1/2"	3/4"	1"	1-1/4"	1-1/2"	2"	2-1/2"	3"	3-1/2"	4"	5"	6"	8"	10"
Solder, Lbs.	1/2	3/4	1	1-1/2	1-3/4	2	2-1/2	3-1/2	4-1/2	5	6-1/2	9	17	35	45
Flux, Oz.	1	1-1/2	2	3	3-1/2	4	4	7	9	10	13	18	34	70	90

Estimating the Cost of Pipe and Fittings. To estimate the cost of pipe and fittings required for any heating plant, obtain the total number of square feet of radiation in the job (based on cast iron radiation) and figure at $2.00 to $2.15 per square foot. This is based on prices of pipe as given on the previous pages.

For one-story residences having short pipe runs, figure the cost of the pipe and fittings at $2.00 per square foot of radiation (based on standard cast iron radiation).

An example of the method used in estimating the cost of pipe and fittings is as follows: for a job containing 1,400 square feet of standard radiation, take 120% of 1,400, which is 1680. The pipe and fittings will cost approximately $3,400.00.

The above allowances for pipe and fittings do not include items such as radiator valves, floor and ceiling plates, boiler fittings and expansion tanks, but only the pipe and fittings required for roughing-in the job.

LABOR INSTALLING HEATING PLANTS

The labor cost of installing steam, hot water, vapor and vacuum heating systems is usually figured on the basis of the number of radiators or convectors (commonly termed units) in the job and estimated at a certain price per unit. This allowance includes all necessary labor to rough-in the job, set and connect boiler, set and connect radiators, convectors or baseboard radiation and complete the job ready to operate. This allowance varies with the type of heating system, whether a one- or two-pipe system and the class of building.

Labor Installing One-Pipe Heating Systems in One Story Houses. To install one-pipe heating systems in one story houses, it requires a steam fitter about 7 hours per unit (radiator). The labor cost per unit should average as follows:

	Hours	Rate	Total	Rate	Total
Steam Fitter	7.0	--	--	$29.30	$205.10

Labor Installing One-Pipe Heating Systems in Two-Story Residences. When installing one-pipe heating systems in two-story residences, it will require about 10 hours steam fitter time per unit (radiator), and the labor cost per unit should average as follows:

	Hours	Rate	Total	Rate	Total
Steam Fitter or Plumber	10.0	--	--	$29.30	$293.00

Labor Installing One-Pipe Heating Systems in Two- or Three-Story Apartment Building. For this work figure about 8 steam fitter hours per unit (radiator). The labor cost per unit should average as follows:

	Hours	Rate	Total	Rate	Total
Steam Fitter or Plumber	8.0	--	--	$29.30	$234.40

Labor Installing One-Pipe Steam Heating Systems in Stores, Theaters, Public Garages, One-Story Factories and Warehouses. When installing one-pipe steam heating systems in any of these buildings, figure about 16 hours steam fitter time per unit (radiator). The labor cost per unit should average as follows:

	Hours	Rate	Total	Rate	Total
Steam Fitter or Plumber	16.0	--	--	$29.30	$468.80

Labor Installing Two-Pipe Hot Water Heating Systems in One-Story Houses. For this work figure about 14 hours steam fitter time per unit (radiator), and the labor cost per unit should average as follows:

	Hours	Rate	Total	Rate	Total
Steam Fitter or Plumber	14.0	--	--	$29.30	$410.20

Labor Installing Two-Pipe Hot Water Heating Systems in Two- or Three-Story Apartment Buildings. This type of work will require about 14 hours steam fitter time per unit (radiator). The labor cost per unit should average as follows:

	Hours	Rate	Total	Rate	Total
Steam Fitter or Plumber	14.0	--	--	$29.30	$410.20

Labor Installing Two-Pipe Hot Water Systems in Residences and Other Types of Buildings. When installing two-pipe heating systems in residences or other types of buildings not specifically mentioned, figure 16 hours steam fitter time per unit. The labor cost per unit should average as follows:

	Hours	Rate	Total	Rate	Total
Steam Fitter or Plumber	16.0	--	--	$29.30	$468.80

Chapter 12

ROOFING, GUTTERS AND DOWNSPOUTS

The demand for conventional asphalt shingles continues, but fiberglass asphalt shingles have become the leading product in the industry. Availability of raw materials and manufacturing control are two main reasons. The supply of basic raw material, silica or sand, is virtually unlimited. The cost of silica is relatively stable when compared with the price of the wood fiber and paper used in organic mat shingles.

In addition to requiring less energy, the manufacturing process for the fiberglass product is more consistent, assuring a uniform product. For the remodeling contractor this means three things: fiberglass shingles are 75% more weatherproof than organic mat shingles; warranties are 5 to 15 years longer than for most conventional shingles; and fiberglass shingles carry an A fire rating instead of C.

Fiberglass shingles for a long time were confined to the sunbelt, because they became brittle as they aged and could not stand up to more severe northern conditions. But new manufacturing processes eliminated the brittleness, and the use of this shingle type moved northward.

For the high-quality remodeling job, cedar shingles are still available. They can be treated with fire-retardant chemicals and carry a UL class C fire rating. And they are moving toward B. Now, however, wherever a C rating exists, contractors have the option of offering their more affluent clients a higher quality product.

Selecting Materials for Re-Roofing. It is best to use the same type of materials as was originally used on the roof. If it is necessary to use a different type for re-roofing, then a number of factors should be considered. Will the new material be as weatherproof and compatible with the pitch of the existing roof? Is the color and pattern compatible with other architectural features? Can the existing framing support the new material? How do installation and costs compare?

Any material can be used on a high-pitched roof. Roofs lower than 1/2 pitch limit the choice of materials. In general wooden shingles and shakes are seldom used for a pitch less than 1/2. For a pitch of 1/6, asphalt shingles are generally used, because they lie flatter than wooden shingles.

A popular option for re-roofing over asphalt shingle is galvanized metal roofing. Metal panels can be applied directly over existing shingles. This type of roofing is available in a number of attractive colors and offers the same warranties and certificates of performance as more conventional types of residential roofing.

ROOFING SHINGLES AND TILES

Roofing is estimated by the *square*, which contains 100 sq. ft. The method used to compute the quantities will vary with the kind of roofing and shape of roof. The labor cost of applying any type of roofing will be governed by the pitch or slope of the roof, size, plan (whether cut up with openings, such as skylights, penthouses, gables and dormers) and on the distance of the roof above the ground.

Rules for Measuring Plain Double Pitch or Gable Roofs. To obtain the area of a plain double pitch or gable roof as shown in Figure 1, multiply the length of the ridge (A to B), by the length of the rafter (A to C). This will give the area of one-half the roof. Multiply this by 2 to obtain the total sq. ft. of roof surface.

Example: Assume the length of the ridge (A to B) is 30'-0" and the length of the rafter (A to C) is 20'-0"; 30 x 20 = 600 sq. ft.; 600 x 2 (sides of roof) = 1,200 sq. ft. of roof area.

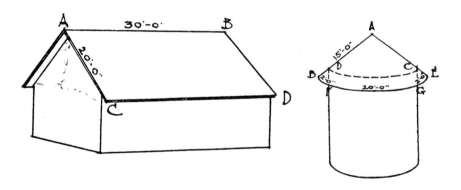

Fig 1. Plain Double Pitch or Gable Roof   Fig 2. Conical Building and Roof

Rules for Measuring Hip Roofs. To obtain the area of a hip roof as shown in Figure 2, multiply the length of the eaves (C to D) by 1/2 the length of the rafter (A to E). This will give the number of square feet of one end of the roof, which multiplied by 2 gives the area of both ends. To obtain the area of the side of the roof, add the length of the ridge (A to B) to the length of the eaves (D to H). Divide the sum by 2 and multiply by the length of the rafter (F to G). This gives the area of one side of the roof and when multiplied by 2 gives the number of sq. ft. on both sides of the roof. To obtain the total number of sq. ft. of roof surface, add the area of the two ends to the area of the two sides. This total divided by 100 equals the number of squares in the roof.

Example: Assume the length of the eaves (C to D) is 20'-0" and the length of the rafter (A to E) is 20'-0". Multiply (C to D) 20'-0" by 1/2 of the length of the rafter (A to E), or 10'-0", which equals 200 sq. ft., the area of one end of the roof. To obtain the area of both ends, 200 x 2 = 400 sq. ft.

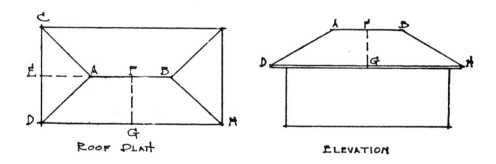

Fig 3. Hip Roof

To obtain the area of the sides of the roof, take the length of the ridge (A to B) which is 10'-0", and the length of the eaves (D to H) which is 30'-0". The combined length of ridge and eaves is 40'-0". Take 1/2 of 40'-0" for the average roof length, 20'-0". Assuming the length of the rafter (F to G) is 20'-0", 20 x 20 = 400 sq. ft. (area of one side of roof). 400 x 2 = 800 sq. ft. (area of both sides of roof). Adding the area of the two ends and the area of the two sides equals 1,200 sq. ft. of roof area.

The area of a plain hip roof running to a point at the top, as shown in Figure 3, is obtained by multiplying the length of the eaves (B to E) by 1/2 the length of the rafter (A to F). This gives the area of one end of the roof. To obtain the area of all four sides, multiply by 4.

Example: Multiply the length of the eaves (B to E), which is 30'-0", by the length of the rafter (A to F), 10'-0". The result is 300 sq. ft., the area of one end of the roof. 300 x 4 = 1,200 sq. ft., the total area for all four sides.

Rules for Measuring Conical Tower Roofs and Circular Buildings. To obtain the area of a conical tower roof, as shown in Figure 4, multiply 1/2 the length of the rafter (A to B) by the distance around the eaves at B. For example, assume the length of the rafter (distance from A to B) is 15'-0"and the diameter of the building at B is 20'-0".

To obtain the distance around the building, multiply the diameter by 3-1/7 or 3.1416. If the eaves project beyond the outside walls, and the diameter is given only to the outside walls of the building, add the length of the roof projection on both sides of the building to obtain the correct diameter. Example: If the diameter of the building (C to D) is 20'-0" and the eaves project 2'-0" on each side, the diameter of the building at the eaves would be 24'-0".

Multiply 24'-0" (B to E), the diameter of the building at eaves, by 3.1416, which equals 75.3984, or approximately 75'-5" around the eaves at projection (B to E).

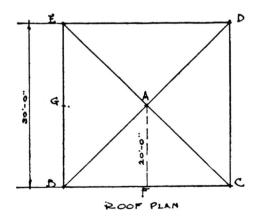
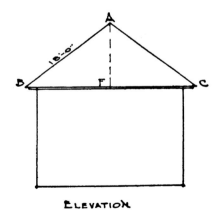

Fig 4. Hip Roof

ROOFING, GUTTERS AND DOWNSPOUTS

To obtain the area of the roof, multiply 1/2 the length of the rafter (A to B), which is 15'-0", (1/2 of 15 is 7-1/2) or 7'-6" by the distance around the eaves at B and E, which is 75'-5" or 75.4' and the result is 565.5 or 565 1/2 sq. ft., the area of the roof.

To obtain the area of a cylindrical or circular building, multiply the height by the circumference, or the distance around the building, and the result will be the number of sq. ft. to be covered. The circumference is obtained by multiplying the diameter (C to D) of the building by 3-1/7 or 3.1416. To obtain the area of the outside walls of a cylindrical building whose diameter is 20'-0" and the height 15'-0", 20 x 3.1416 = 62.832 or 62'-10". Multiply 62.832 (the distance around the building) by 15'-0" (the height of the building) = 942.48 or 942 1/2 sq. ft., the area of the outside walls.

A SHORT METHOD OF FIGURING ROOF AREA

To obtain the number of square feet of roof area, where the pitch (rise and run) of the roof is known, take the entire flat or horizontal area of the roof and multiply by the factor given below for the roof slope. The result will be the area of the roof. Always bear in mind that the width of any overhanging cornice must be added to the building area to obtain the total area to be covered.

Example: Find the area of a roof 26'-0" x 42'-0", having a 12" or 1'-0" overhanging cornice, a 1/4 pitch or a 6 in 12 rise and run. To obtain the roof area, 26'-0" + 1'-0" + 1'-0" = 28'-0" wide. 42'-0" + 1'-0" + 1'-0" = 44'-0" long. 28 x 44 = 1232 or 1,232 sq. ft. flat or horizontal area.

To obtain the area at 1/4 pitch, or 6 in 12 rise and run, multiply 1,232 by 1.118 = 1377.376 or 1,378 sq. ft. of roof surface. Add allowance for overhang on dormer roofs and sides.

Pitch of Roof	Rise and Run	Multiply Flat Area by	Lin. Ft. of Hips or Valleys per Lin. Ft. of Common Run
1/12	2 in 12	1.014	1.424
1/8	3 in 12	1.031	1.436
1/6	4 in 12	1.054	1.453
5/24	5 in 12	1.083	1.474
1/4	6 in 12	1.118	1.500
7/24	7 in 12	1.158	1.530
1/3	8 in 12	1.202	1.564
3/8	9 in 12	1.250	1.600

Pitch of Roof	Rise and Run	Multiply Flat Area by	Lin. Ft. of Hips or Valleys per Lin. Ft. of Common Run
5/12	10 in 12	1.302	1.612
11/24	11 in 12	1.357	1.685
1/2	12 in 12	1.413	1.732

Hips and Valleys. The length of hips and valleys, formed by intersecting roof surfaces, running perpendicular to each other and having the same slope is also a function of the roof rise and run. For full hips or valleys, i.e. where both roofs intersect for their full width, the length may be determined by taking the square root of the sum of the rise squared plus twice the run squared.

Using the factors given in the last column of the above table, the length of full hips or valleys may be obtained by multiplying the total roof run from eave to ridge, (not the hip or valley run), by the factor listed for the roof slope involved.

Your total estimate should include the following: the cost of the roofing materials and underlayment, nails, roof coatings, plastic roof cements, flashing materials, equipment rentals, replacing any boards at the edge of the roof and repair, if required, of sheathing. *Warning: keep clear of electrical power lines that attach to the roof.*

Removing Roofing Materials. Take care not to damage flashing on chimneys, vents and valleys. Use the old flashing as a pattern to cut the new. To remove asphalt shingles, figure one roofer and one laborer working together can strip around 20 squares per 8-hour day. Add 2 hours to remove debris and clean up. To remove eaves flashing and drip edges, figure one roofer can remove 50 lin. ft. per hour. To remove chimney and vent flashing, figure one roofer can remove and clean it in preparation for replacement in 3 hours for a chimney and one hour for a vent pipe. Add one hour to remove debris and clean up.

PREPARATION FOR NEW ROOFING

Installing Underlayment. *Note: Do not use underlayment for wooden or shake shingle roof. Moisture trapped between the underlayment and the sheathing will cause the shingles to decay or be attacked by fungus.* A roofer and laborer should be able to install 2500 sq. ft. of underlayment per 8-hour day at the following cost per 100 sq. ft.:

ROOFING, GUTTERS AND DOWNSPOUTS

	Hours	Rate	Total	Rate	Total
Roofer	0.32	--	--	$24.60	$ 7.87
Laborer	0.32	--	--	17.95	<u>5.74</u>
Cost per 100 s.f.			--		13.61
Cost per s.f.			--		0.14

If using 15-lb. asphalt saturated roofing paper as underlayment, figure $11.95 per roll (432 sq. ft. roll).

To Install Drip Edges. Drip edges are usually made of aluminum, galvanized steel or copper. Use nails of the same materials to prevent corrosion. Space nails 1" from edge and no more than 10" apart. Where two drip edges meet, seal with asphalt roofing cement. Material costs are as follows: aluminum, $0.30 per ft.; galvanized steel, $0.28 per ft.; copper, $1.50 per ft. A roofer should be able to install 60 lin. ft. per hour.

To Install Eaves Flashing. Eaves flashing strips are made from 65- or 90-lb. mineral surfaced roll roofing material. The 90-lb. material is preferred, because it will last longer. In areas of moderate to heavy snowfall, install a double thickness. Material costs are $12.55 for 50-lb. material and $14.66 for 90-lb. material. Figure one roofer can install 60 lin. ft. per hour.

Single Thickness. Install with a minimum of 6" head lap until the top of the eaves flashing is at least 12" (24" if pitch is 1/6 or less) past the inside wall line of the house. Space nails 1" from top edge and 24" apart. Bend flashing to overhang. Apply roofing lap-cement and nail flashing to roof. Start 1" from drip edges and space nails 4" apart. Apply lap-cement to nail heads. One roofer should be able to install 60 lin. ft. per hour at a cost of $0.41 per lin. ft.

Installing Chimney Flashing. Using the old pieces as patterns, a local sheet metal shop can cut and shape new flashing. Nail flashing to sheathing, taking care to use nails of the same material as the sheathing. Apply a clear butyl or aluminized compound to the back of the cap flashing and press into place. Caulk all edges of cap flashing and mortar joints. Figure one roofer can replace 15 lin. ft. per hour, providing a satisfactory raggle exists. If a raggle has to be chiseled, figure one laborer can cut 8 to 10 lin. ft. per hour. With an electric saw, figure 12 to 15 lin. ft. per hour.

Chimney Flashing Table

No. of Brick Wide	No. of Brick Long	Inches Wide	Inches Long	Pitch of Roof to 1/3	Pitch of Roof to 1/2	Pitch of Roof to 3/4
				Lin. Ft. of Flashing		
2	2	18"	18"	11'	11'	12'
2	2-1/2	18"	22"	12'	12'	13'
2	3	18"	26"	13'	13'	15'
2	3-1/2	18"	30"	13'	14'	16'
2	4	18"	34"	14'	15'	17'

Installing Valley Flashings. *Composition Flashings.* Composition roofing valley flashing is constructed the same way for both open and closed valley flashings. *Note: Do not leave strips of roofing on a lawn in the sun. Absorbed heat will burn the lawn.* Use old flashing as a pattern. Cut an 18" width of material to fit valley, allowing 4" to 7" for overhang of drip edges. Place the strip mineral side down, applying roofing lap-cement to the underside of the vee at the drip edge and nail sheathing 1" from edge of material 12" apart.

Using the same pattern, cut a 36" wide strip of material 30" longer than the valley centerline to fit the valley. Cut a vee using the old pattern and place the mineral side face up on the 18" strip. Fasten with roofing lap-cement and nails. Figure one roofer can install 40 lin. ft. per hour at a cost of $0.62 per lin. ft.

Metal Flashing. Metal flashing is installed the same way as open or closed valley flashing except that it is usually installed over 30- or 65-lb. composition roll roofing. With metal flashing when the pitch is 1/2 or more, the flashing must extend at least 7" on each side of the valley centerline. If less than 1/2 pitch, it must extend at least 10" per side.

Installing Vent Pipe Flashings. These can be purchased from most building material suppliers in ready-to-install form. Vent pipe flashings should have flanges that extend 8" on top of the vent, 6" on each side and 4" below. Figure one worker can install such flashings at a rate of 0.5 hours per unit.

Flashing Between Roof and Vertical Walls. These flashings are made from 65- or 90-lb. composition material. If the vertical wall has lap siding, it is best to use flashing strips 12" long bent to slide 4" up under the siding. Since these strips are installed as roofing is applied, add 10 minutes labor time per course.

It is best not to use asphalt roofing cement on the side of flashing strips that will come in contact with wood or shake shingles. The wood will draw the oil out of the cement and cause the shingles to warp or split.

Flashing Vertical Walls Without Lap Siding. Cut an 8" wide strip of 90-lb. composition roofing material equal to the length to be flashed. Apply asphalt roofing cement to the underside of the strip and place it flush against the vertical wall. Nail it to the wall, spacing the nails 1" from edge and 10" apart, and coat the nail heads with roofing cement. Apply a thick bead of roofing cement to the edge of flashing against vertical wall. Figure one worker can apply 40 lin. ft. at a cost of $0.62 per lin. ft.

INSTALLING NEW ROOFING

Asphalt shingles come in several different styles. Strip shingles are furnished in three tab square butt strips, two tab hex strips and three tab hex strips. Installation is the same for all three, except when installing shingles as a starter course and starting subsequent courses. Individual asphalt shingles are furnished and laid in exactly the same manner as wood shingles.

Estimating Quantities of Asphalt Shingles. When measuring roofs of any shape, always allow one extra course of shingles for "starters" at the eaves. The first course must always be doubled. An alternative is to use composition roll roofing as the starter course.

Measure the length of all hips, valleys and ridges, figuring each at 1'-0" wide. Multiply and add these amounts to the total for the roof. Many roofing contractors do not measure hips, valleys and ridges. Instead they prefer to add a predetermined percentage to cover material and waste. Percentages commonly used are: gable roof, 10%; hip roof, 15%; hip roof with dormers and valley, 20%.

High winds can blow asphalt shingles loose, if they are not properly nailed. Six nails to the strip, nailed right to the cutout, is recommended. In addition to self-sealing shingles, most manufacturers produce a shingle designed for high wind areas. They interlock in such a manner that all shingles are integrated into a single unit. Interlocking shingles are available in single coverage for reroofing and double coverage for new construction. When using asphalt shingles for roofs, roll-asphalt roofing of the same material is often for forming valleys, hips and ridges.

Nails Required for Asphalt Shingles. When laying individual asphalt shingles, use 12-ga. galvanized nails, 1-1/2" long with 7/16" heads. For laying over old roofs, use 1-1/4" nails. When laying square butt strip shingles, use 11-ga. galvanized nails 1" long, and when laying over old roofs, use 1-1/4" long nails.

Sizes and Estimating Data for Asphalt Shingles

Kind of Shingle	Size	No. Shingles per Sq.	Expos. Inches	Length Nails	No. Nails per Shingle	Lbs. Nails per Sq.
3-in-1 strip	12" x 36"	80	5	1	4	1
3-in-1 strip	15" x 36"	80	5	1	4	1

Approximate Prices of Mineral Surfaced Asphalt Shingles

Kind of Shingle	Size	Weight per Square	Price per Square
3-in-1 strip	12" x 36"	235 lbs.	$26.00
3-in-1 strip	12" x 36"	300 lbs.	37.00
3-in-1 strip	15" x 36"	325 lbs.	52.00
3-in-1 strip	15" x 36"	350 lbs.	59.00

Most asphalt shingles carry an Underwriters Class "C" rating. With fiberglass an "A" rating can be achieved. Class "A" 235-lb. shingles are approximately $35.00 per square, and 300-lb. are $65.00 per square.

Starter Course Using Composition Roll Roofing. Cut the starter course from 90-lb. roll roofing, 12" wide and 3/4" longer than length of roof. Place it face down, and allow the ends to extend 3/8" beyond the gables and eaves. Secure this starter course with asphalt cement and nail it, spacing nails 8" apart and 1" in from the top and sides. One roofer can do this work at a rate of 40 lin. ft. per hour.

Installing Three Tab Butt Strip Shingles. When the starter course is a row of shingles, begin with a partial shingle strip. Place the starter course shingles face down, allowing 3/8" overhang at gable and eaves. Nail through the drip edge. Continue the starter course using full-size strips, four nails per strip.

Install the first course over the starter course, placing a whole strip at a gable corner, mineral side up and tabs down. Place daubs of cement over nail heads and press the strip into position. *Note: Do not heat asphalt roofing cement over fire. Explosion or fire could result. An acceptable method is to place can in warm water.*

Nail strip in place with six galvanized nails on a line just above the ends of the shingle slots. Place subsequent courses centering the tabs so that alternating courses are in alignment. A roofer can lay one square of asphalt shingles in two hours on double pitched roofs with no hips, valleys or dormers. Figure about 35

ROOFING, GUTTERS AND DOWNSPOUTS

lin. ft. per hour for fitting hips, valleys and ridges. Hip and ridge material is usually cut on the job from regular composition roll roofing.

Estimating Asphalt Roofing Materials

Roofing Material	Shingles per Sq.	Nails per Shingle	Nail Length	Nails per Sq.	Estimated Lbs. per Square 12 ga. 7/16" Hd.	Estimated Lbs. per Square 11 ga. 7/16" Hd.
Roll Roofing (new)			1"	252*	0.73	1.12
Roll Roofing (over old)			1-3/4"	252*	1.13	1.78
3-Tab Sq. Butt (new)	80	4	1-1/4"	336	1.22	1.44
3-Tab Sq. Butt (over old)	80	4	1-3/4"	504	2.38	3.01
Hex Strip (new)	86	4	1-1/4"	361	1.28	1.26
Hex Strip (over old)	86	4	1-3/4"	361	1.65	2.03
Giant American	226	4	1-1/4"	479	1.79	2.27
Giant Dutch Lap	113	2	1-1/4"	236	1.07	1.39
Single Hex	82	2	1-3/4"	172	0.79	1.03

*Spaced 2" apart.

Asphalt Shingles
Dimensions, Weights and Exposures

Roofing Material	Length	Width	Packages per Square	Shingles per Square	Wgt per Sq.	Exposure
2- and 3-Tab Sq. Butt	36"	12"	3	80	235	5"
2- and 3-Tab Hexagonal	36"	1-1/3"	3	86	195	5"
Individual Staple and Individual Lock	16"	16"	2	80	145	
Giant American	16"	12"	4	226	330	5"
Giant Dutch Lap	16"	12"	2	113	165	10"

Recommended Nails/Asphalt Shingles

Installation	1" Sheathing	3/8" Plywood
Re-roofing over asphalt roofing	1-1/2"	1"
Re-roofing over wood shingles	1-3/4"	
New roofing strip or individual shingles	1-1/4"	7/8"

Cedar Shakes
Estimated Coverage per Square

		Estimated Sq. Ft. Coverage Exposure								
Size	Shake	5-1/2"	6-1/2"	7"	7-1/2"	8"	8-1/2"	10"	11-1/2"	13"
18-1/2"x1/2" to 3/4"	Handsplit&Resawn	55*	65	70	75**					
18-3/4"x3/4" to 1-3/4"	Handsplit&Resawn	55*	65	70	75**					
24"x1/2" to 3/4"	Handsplit&Resawn		65	70	75*	80	85	100**		
24"x3/4" to 1-1/4"	Handsplit&Resawn		65	70	75*	80	85	100**		
32"x3/4" to 1-1/4"	Handsplit&Resawn							100**	115	130**
24"x1/2" to 5/8"	Taper-split		65	70	75*	80	85	100**		
18" x 3/8"	St.-split	65*								
24" x 3/8"	St.-split		65	70	75*					

*Recommended for 3-ply construction
**Recommended for 2-ply construction

Cost of 100 Sq. Ft. of 300-lb. Asphalt Shingles on Plain Double Pitch or Gable Roofs

	Hours	Rate	Total	Rate	Total
1 Sq. Strip Shingles		--	--		$37.00
1 Lb. Galvanized Nails		--	--		0.75
Roofer		--	--	$21.25	24.60
Cost per 100 sq. ft.			--		$62.35
Cost per sq. ft.			--		0.62

On roofs with gables and dormers, add 15 minutes per square. On more complex hip or English type roofs, add 12 minutes per square. If roofers are experienced, deduct 10 minutes per square.

ROOFING, GUTTERS AND DOWNSPOUTS

Cost of 100 Sq. Ft. of 325-lb. Asphalt Shingles on Plain Double Pitch or Gable Roofs

	Hours	Rate	Total	Rate	Total
1 Sq. Strip Shingles	--	--	--		$52.00
1 Lb. Roofing Nails	--	--	--		0.75
Roofer	--	--	--	$21.25	24.60
Cost per 100 sq. ft.			--		$77.35
Cost per sq. ft.			--		0.77

On roofs with gables and dormers, add 12 minutes per square. On difficult hip or English type roofs, add 10 minutes per square. If roofers are experienced, deduct 15 minutes per square.

Roofing Material Costs

Roll Roofing
90-lb. Slate Coat (100 sq. ft. coverage - 4 colors) ....$14.66 roll
50-lb. Smooth (100 sq. ft. coverage) ....12.55 roll
Selvedge Edge (50 sq. ft. coverage) ....12.33 roll
15-lb. Felt (432 sq. ft. coverage) ....13.88 roll
30-lb. Felt (216 sq. ft. coverage) ....13.88 roll
9" Starter (36') ....5.22 roll
18" Valley (36') ....9.33 roll

Aluminum Flashing
14" x 10" $7.19
14" x 50" 31.55
20" x 10" 9.77
20" x 50" 30.95
5 x 7 Step Shingles ....0.14 ea.

Roof Edge (10' Aluminum)
White ....2.39 ea.
Brown ....2.99 ea.

Rake Edge (10')
White 2.99

Plastic Roof Cement
One Gallon 4.29
Five Gallon 16.55

WOOD SHINGLES AND SHAKES

Wood shingles are still popular in localities throughout the country where building codes permit their use. They are a premium product, and volatile prices are often the deciding factor in the decision to go with a different roofing material instead. There are several types of composition shingle that are made to look like shakes with surprisingly good results. The customer can have a very attractive roof at a substantially lower price.

Wood shingles are sold by the square, which covers 100 sq. ft. and generally considered to contain 1,000 shingles that are 4" wide. Widths range from 4" to 14", and widths greater than 8" are usually halved and used as separate shingles. When re-roofing, always use the same shingles and exposure. The following exposure guidelines are standard:

Standard Shingle Exposures

Single Length	1/5 Pitch or More	1/5 Pitch or Less
16"	5"	3-3/4"
18"	5-1/2"	4-1/4"
24"	7-1/2"	5-3/4"

The labor cost of laying wood shingles depends on the type of roof, whether a plain gable roof, a steep roof or one cut up with gables and dormers, and on the manner in which shingles are laid— regular butts, irregular or staggered butts or thatched butts.

The costs given on the following pages are based on the actual number of shingles a roofer lays per day and not on the number of squares, which varies with the spacing of the shingles. It does not make any particular difference to the roofer whether the shingles are laid 4", 4-1/2" or 5" to the weather. He will lay practically the same number no matter what the spacing. It does make considerable difference in the number of sq. ft. of surface covered, which will vary from 10% to 40%.

ROOFING, GUTTERS AND DOWNSPOUTS

The number of shingles laid in a given amount time does vary with the ability of the roofers and the class of work. Ordinary carpenters cannot lay shingles as fast and well as carpenters who specialize in shingle laying. Of course, experienced shinglers will demand a higher wage rate for their work.

Some carpenters claim to be able to lay 16 bundles (3,200 shingles) per 8-hour day, but this is unusual and is found only on the cheapest grade of work, where only one nail is driven into each shingle instead of the two needed to secure a workmanlike job.

Estimating the Quantity of Wood Shingles. Ordinary wood shingles are furnished in random widths, but dimension shingles are sawed to uniform widths of 4", 5" and 6". Wood shingles are sold by the square with sufficient shingles to lay 100 sq. ft. of surface, when laid 5" to the weather. There are four bundles per square.

When estimating the quantity of ordinary wood shingles required to cover any roof, bear in mind that the distance the shingles are laid to the weather makes considerable difference in the actual quantity required. There are 144 sq. in. in 1 sq. ft. and an ordinary shingle is 4" wide. When laid 4" exposed to the weather, each shingle covers 16 sq. in. or it requires 9 shingles per sq. ft. of surface. There are 100 sq. ft. in a square. 100 x 9 = 900, and allowing 10% to cover the double row of shingles at the eaves, waste in cutting and narrowing shingles, it will require 990 shingles (5 bundles) per 100 sq. ft. of surface.

Number of Shingles and Quantity of Nails Required per 100 Sq. Ft.

Distance Laid to Weather	Area Covered by One Shingle Sq. In.	Add for Waste Percent	Actual No. per Square without Waste	Number per Square with Waste	No. of 4-Sq. Bdls. Required	Lbs. 3d Nails Req'd.
4"	16	10	900	990	5.0	3.2
4-1/4"	17	10	850	935	4.7	2.8
4-1/2"	18	10	800	880	4.4	2.5
5"	20	10	720	792	4.0	2.0
5-1/2"	22	10	655	720	3.6	1.6
6"	24	10	600	660	3.3	1.5

Applying Shingles for Different Roof Slopes. Roof pitch is the ratio of a given distance of roof slope over a given distance of run. It is expressed as a fraction, such as 1/8, 1/3 and 1/2 pitch. In the cross section below, the length of distances AB and BC constitutes pitch. The length of AC, which extends from one eave-line to the other, is known as the *span*. The length of AD or DC is one-half of the span and is called the *run,* and distance BD is called the *rise.* The relationship of the rise to run obviously is related to the slope of AB or BC. In fact, roof pitches are computed from the ratio of rise to run. Therefore, the first step is to determine length of the run (AD or DC) and the rise (BD).

Wood shingles are manufactured in three lengths—16", 18" and 24". The standard weather exposure (portion of shingle exposed to weather) for 16" shingles is 5", for 18" is 5-1/2" and for 24" shingles is 7-1/2". These standard exposures are recommended on all roofs of 1/4 pitch and steeper (6" rise in 12" run). On flatter roof slopes the weather exposure should be reduced to 3-3/4" for 16" shingles, 4-1/4" for 18" shingles and 5-3/4" for 24" shingles.

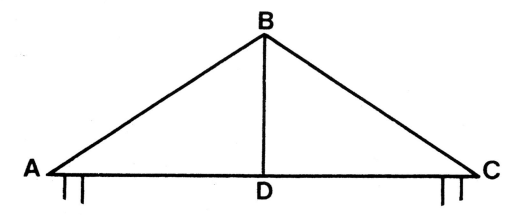

This diagram shows at a glance the weather exposure to be used for various roof pitches. For example, if a roof has a rise of 8" in a run of 12", it can be seen that this is 1/3 pitch and that an exposure of either 5", 5-1/2" or 7-1/2" should be employed, depending on the length of the shingles used.

Installing Wooden Shingles. A starter shingle is placed at a corner where gable and eave meet, overhanging 3/4" at gable and extend 1-1/2" beyond sheathing. Nails should be 1 to 1-1/2" on the gable side. *Note: Be sure nails do not go through sheathing. Nails should be flush with surface. Do not crush shingle. Early decay can result.*

Install a second shingle at other end of the course. Nail a chalk line to the butt ends of these end shingles that you have placed and nail several shingles, aligned with chalk line, randomly in between the end shingles. Use two nails per shingle, each nailed 3/4" from a side of the shingle. Install the remaining shingles by aligning with the placed shingles, allowing 1/4" between shingles to permit swelling when wet.

ROOFING, GUTTERS AND DOWNSPOUTS

The first course of shingles is placed on the starter course so that the gaps between shingles are offset from gaps in the starter course by 1-1/2". Nail using two nails per shingle. A roofer and helper can install 4 to 6 squares of wooden shingles per 8-hour day at the following cost per 100 sq. ft.:

	Hours	Rate	Total	Rate	Total
1 Sq. Wooden Shingles	--	--	--	$80.00	$ 80.00
4 Lbs. Nails	--	--	--	3.00	3.00
Roofer	1.3	--	--	24.60	31.98
Helper	1.3	--	--	17.95	23.34
Cost per 100 s.f.			--		$138.32

Do not install shingles in valleys. Go only to the valley chalk line. For cutting and fitting shingles next to a valley, figure one roofer can install 20 lin. ft. per hour. Hip and ridge shingles can be bought prefabricated. Their length should be twice the exposure. One roofer can usually lay 35 lin. ft. per hour.

Installing Wooden Shakes. Shakes are installed the same as shingles with the following exceptions. Shingles are used as the starter course for shakes. Thus you should order enough shingles for a starter course and add this cost to your estimate. Since shakes are hand split, the sides will be uneven, and the gap between shingles should be about 1/4" to 3/8". Shakes are applied over 1" x 4" roof sheathing that requires an underlayment of roofing paper.

The first shake in a course at a gable edge should extend 1" beyond the edge. An additional labor factor is that shakes call for an underlayment between the first and succeeding courses. Use an 18" wide strip of 30-lb. roofing paper covering the top of the shakes and sheathing above. The paper is placed twice the exposure from the butt and parallel to butt ends. Add 0.3 hour per square for roofer and 0.3 hours per helper for placing this material.

ROLL ROOFING

Composition roll roofing material is 3' wide and 36' long, designed to cover 100 sq. ft. of roof with 6" vertical and horizontal laps. Standard rolls weigh 87 lbs. and are called 90-lb. roofing. Underlayment and the eaves flashing strip are not required with roll roofing.

Roll roofing can be installed horizontally or vertically with nails exposed or concealed. On roofs with a pitch of 7/12 or steeper, it should be installed horizontally; on lower pitch roofs, it can be installed vertically.

Roll roofing should be applied in warm weather. In cool weather it can crack. Install it on dry surfaces only, or otherwise, it can decay. Always avoid walking on it. Rolls left in the sun on a lawn can absorb sufficient heat to burn the grass.

Installing Roll Roofing Vertically. Measure from the eave to the ridge, allowing an extra 4-3/4" for overhang at eaves and ridges. Place a strip parallel to gable, extending 4" beyond ridge and 3/4" beyond eave. Tack the strip in place and nail using 7/8" galvanized roofing nails 1" in from the edge and 2" apart. Mark sheathing cracks lightly and space nails in next strip to avoid the cracks. Overlap the next strip 6" and coat lap with roofing lap cement and nail vertical lap 4" apart, 1" in from the edge. Nail the bottom with nails spaced 2" apart and 1" up from the eave. Succeeding strips are installed in the same manner. The 4" extension at the ridge is bent over the ridge and nailed to sheathing, spacing nails 8" apart and 1" up from the end of the strip. At the gables and eaves, nail to the sheathing with nails 6" apart. Apply molding strips to overhang of both eaves and gables. A roofer and helper should be able to install 1,900 sq. ft. of roll roofing per day vertically at the following cost per 100 sq. ft.:

	Hours	Rate	Total	Rate	Total
1 Sq. Roll Roofing		--	--		$ 14.66
1 Lb. Nails		--	--		0.75
Roofer	0.4	--	--	$24.60	9.84
Helper	0.4	--	--	14.10	7.18
Cost per 100 s.f.			--		$32.43
Cost per s.f.			--		0.32

If a course ends at a valley, add 20 minutes per strip. If the course ends at a hip, add 20 minutes per strip. See section below for ridges.

Installing Roll Roofing Horizontally. The procedure for applying roll roofing horizontally is about the same as installing it vertically, but it tends to take longer because of the difficulty working on a steeper pitch. Figure a two-worker crew, roofer and helper, can lay 6 to 8 squares of horizontal roofing per 8-hour day at the following cost per 100 sq. ft.:

	Hours	Rate	Total	Rate	Total
1 Sq. Roll Roofing		--	--		$14.66
1 Lb. Nails		--	--		0.75
Roofer	0.5	--	--	$24.60	12.30
Helper	0.5	--	--	14.10	8.98
Cost per 100 s.f.			--		$36.69
Cost per s.f.			--		0.37

Hips and Ridges. To finish hips and ridges, figure one worker can lay 100 lin. ft. of roll roofing material per hour.

GALVANIZED METAL ROOFING

Metal roofing is becoming popular in the U.S., especially in the sunbelt, because of the advantages it offers in durability and energy-savings. Unlike asphalt and wood shingles, metal roofing reflects heat rather than absorbing it.

Metal costs about twice as much as conventional shingle, but the higher price is offset by savings that can be gained in a number of ways. There is no need for plywood under the roofing, and in fact, for a re-roofing job, the light-weight metal panels can go on over existing shingles. Of course, the contractor will need to install insulation and vapor barrier because of the potential for condensation.

This type of roofing comes in lengths up to 45', requiring two workers to handle. There are two types of fastening method. With the *exposed fastener* method, fasteners are driven through pre-formed panel surfaces and panel laps to secure them to the decking or framing. With the *concealed fastener* system, panels are installed over decking using metal clips.

Metal roofing for residential applications is offered in a variety of colors, with up to 25-year warranties and certificates of performance.

GUTTERS AND DOWNSPOUTS

Gutters are made from galvanized steel, aluminum and vinyl. Each comes finished or unfinished. Those with baked finish are offered in a range of colors that can match or compliment the existing siding. The advantages of such finishes, in addition to complimenting the decor, is that they need little or no maintenance and have a considerably longer life span than unfinished products. On the other hand, unfinished products can be painted to match the exact color of the house. Unprotected galvanized steel gutters and downspouts are inexpensive. However, unless they are primed and painted they will rust and corrode. Aluminum is more expensive, but it forms its own protective coating, giving it a considerably longer service span.

Typical Components
1. End Cap
2. Outside Miter
3. End Piece with Outlet
4. Gutter Section
5. Slip Joint Connector
6. Inside Miter
7A. Elbow-Style A
7B. Elbow-Style B
8. Downspout or Conductor Pipe
9. Pipe Strap
10. Expansion Joint

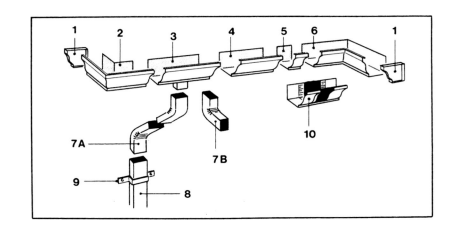

Typical gutter and downspout components.

Because it is lighter, aluminum is easier to install, but it lacks the rigidity of steel, and greater care must be used in handling it or dents and nicks will occur. Vinyl gutters and downspouts, made from polyvinyl chloride (PVC), require very little maintenance but expand and contract as temperatures change. Unless an allowance is made for these physical changes, they can bend and pull away from the roof.

Approximate Material Costs
Gutters, Downspouts and Hardware*

Unit		Minimum Quantity	Aluminum
Gutters	.027/5"x20"	1	0.87 ft.
	.032/6" custom	1	1.30 ft.
	.032/6"x20"	1	1.20 ft.
Downspouts	3"x8'/3"x10'	160	0.46 ft.
		200	
	4"x8'	80	
	4"x10'	100	0.78 ft.
Inside & Outside Miters	5"	20	2.06 ea.
	6"	20	6.12 ea.

ROOFING, GUTTERS AND DOWNSPOUTS

Unit		Minimum Quantity	Aluminum
Strip Miters			1.00 ea.
End Caps	5"	50	0.29 ea.
	6"	50	0.80 ea.
Elbows	3"	30	0.53 ea.
	4"	16	0.93 ea.
Outlets	3"	50	0.38 ea.
	4"	36	0.78 ea.
Pop Rivets		1000	12.80 M
Gutter Sealer		12	1.36 ea.
Gutter Mastic		10	1.15 ea.
Gutter Guard	5"	(75 pcs.) 225 ft.	86.11 pkg
Ferrules	5"	500 ft.	0.05 ea.
	6"	400 ft.	0.06 ea.
Spikes	7"	500 ft.	0.11 ea.
	8"	500 ft.	0.13 ea.
Pipe Bands		100 ft.	0.19 ea.
Drip Edge	3"	500 ft.	0.24 ft.

Custom Gutters

Gutters	$3.04 lin. ft.
Downspout	3.04 lin. ft.
4" Downspout	4.00 lin. ft.
Bay Cuts	8.00 each cut
Tie-Ins—Funnels	12.00
Gutter Guards (Labor & Material)	0.96

Divertors   3" 28.80
            4" 33.60
Each labor accordingly above 30"

Estimated Labor and Materials Combined

Gutters   **Per Lin. Ft.**
8' box for flat roofs $10.40
5" O.G. molded for pitched roof 4.80
For O.G. gutters over 20' above grade add 3.60
5" halfround for pitched roofs 3.60
Over 20' above grade add 3.60
All miters 16.00
Elbows
   3" 3.20
   4" 4.00
   5" 8.00
   6" 9.60
Downspouts
6" for large gutter 5.60
5" for L.G. with halfround 3.20
4" for halfround 2.40
3" for halfround 2.00
Funnels 12.80
Min. charge 160.00

The above prices are for ordinary straight jobs. Any unusual conditions, involving accessibility and height, numerous ends, tubes on large hangers or elbows, must be charged at a higher cost per unit. Add 50% when the work is over 20' high from grade.

Removing Old Gutters and Downspouts. Begin at one end of a gutter run, and work to the other end. Runs of more than six feet require a helper. Try to avoid damage to the existing gutter, because some sections might be reusable, and the old installation can be used as a pattern to install the new material. Remove sheet metal screws in drop outlet, straps and downspouts. For strap hangers, use tin snips to cut straps at the edge of the roof. If bracket hangers cannot be disconnected, also cut with tin snips. Use block and pry bar for spike fasteners. Continue until all gutter sections are removed. Check fascia for damage or

ROOFING, GUTTERS AND DOWNSPOUTS

decay. Fill nail holes and cracks with wood putty. A sheet metal worker and helper should remove 50 lin. ft. per hour at the following cost per 100 lin. ft.:

	Hours	Rate	Total	Rate	Total
Sheet Metal Worker	2.0	--	--	$28.20	$56.40
Helper	2.0	--	--	17.95	35.90
Cost per 100 l.f.			--		$ 92.30
Cost per l.f.			--		0.92

Installing Gutters and Downspouts. Mark the slope of the gutter run, 1/16" to 1/4". Figure a downspout for each corner. However, downspouts are required for every 40' of run, and runs longer than 40' need one at each end.

Cut the gutters to exact length using the old gutters as a pattern and allowing for connectors at each end. Cut unfinished gutters with a hacksaw and enameled gutters with tin snips to avoid chipping. Apply rust-preventive primer to the cut edge.

If using spikes and ferrules, do not use the old spike holes. Using the old gutter as a pattern, locate new holes 1" on either side of old hole. Apply mastic to joints and assemble sections. Attach assembled sections to fascia.

Assemble downspouts using elbows to reach the house wall. Drill two holes at the backside of all joints and fasten with two sheet metal screws or pop rivet. On wood, vinyl or aluminum siding, strap downspouts to wall using 2-penny aluminum or galvanized nails.

A sheet metal worker and helper should attach 1,000 lin. ft. of downspouts and gutters per 8-hour day at the following cost per 100 lin. ft.:

	Hours	Rate	Total	Rate	Total
Sheet Metal Worker	8.0	--	--	$28.20	$225.60
Helper	8.0	--	--	17.95	143.60
Cost per 1,000 l.f.			--		$369.20
Cost per 100 l.f.			--		36.92

To attach downspouts for stucco walls using toggle bolts, add 5 minutes per hour for sheet metal worker. To attach downspouts to brickwalls using lead anchors, add 15 minutes per hour.

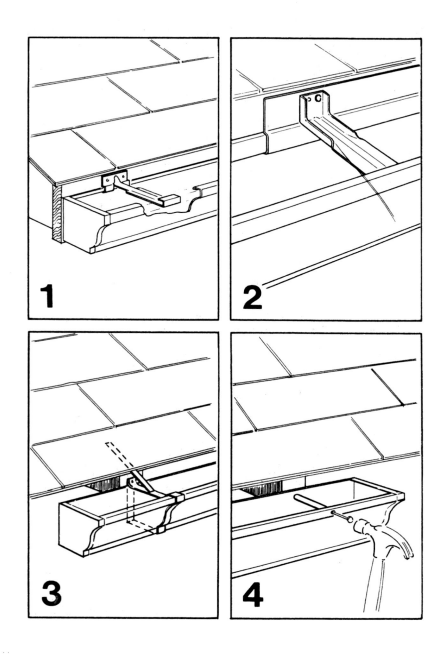

ROOFING, GUTTERS AND DOWNSPOUTS

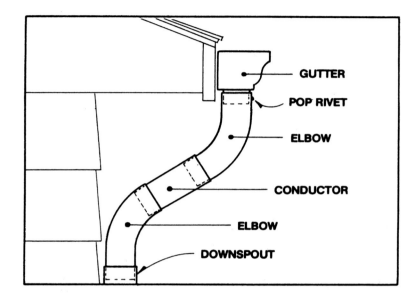

Downspout should be flush with wall for maximum support, and pop-riveted for maximum strength.

QUICK REFERENCE CHART #6
Labor Factors for Insulation

Item	Unit	Factor 16" oc	24" oc
R30 Batt	SF	0.00444	0.00400
R22 Batt	SF		0.00286
R19 Batt	SF	0.00267	0.00235
R11 Batt	SF	0.00235	
1-1/2" CPVB	SF	0.00333	
3-1/2" CPVB	SF	0.00286	
Stick Clip	EA	0.00533	
Weld Pin	EA	0.00533	
Thermostud	SF	0.00667	
on concrete	SF	0.01332	
Styrene			
Glued	SF	0.08000	
Nailed	SF	0.00533	
Batts w/wire			
R3.5 Batt	SF	0.00400	
R5.9 Batt	SF	0.00320	
R11 Batt	SF	0.00267	
R13 Batt	SF	0.00286	
R19 Batt	SF	0.00267	
R30 Batt	SF	0.00533	
R38 Batt	SF	0.00533	

ROOFING, GUTTERS AND DOWNSPOUTS

QUICK REFERENCE CHART #7
Labor Factors for Roofing

Item Description	Unit	Factor New	Reroof
235# Asphalt Seal Tab Shingles			
Plain Gable	SF	0.0111	0.0100
Plain Hip	SF	0.0118	0.0105
Gable w/dormer	SF	0.0125	0.0111
Hip w/dormer	SF	0.0125	0.0111
Gable w/intersecting roof	SF	0.0125	0.0111
Hip w/intersecting roof	SF	0.0125	0.0111
Gable w/dormer & inter. roof	SF	0.0133	0.0118
Hip w/dormer & inter. roof	SF	0.0133	0.0118
Replace shingles	EA		0.0400
Replace tabs	EA		0.0333
90# Mineral Roll Roofing			
Plain Gable - nails exposed	SF	0.0077	0.0077
Plain Hip - nails exposed	SF	0.0083	0.0083
Plain Gable - concealed	SF	0.0091	0.0091
Plain Hip - nails concealed	SF	0.0099	0.0099
Wood Shingles on plain gable or hip w/ridge, hip and valleys included			
16" w/5" exposure	SF	0.0250	
18" w/5" exposure	SF	0.0250	
24" w/5-3/4" exposure	SF	0.0217	
16" w/4-1/2" exposure	SF	0.0278	
18" w/4-1/2" exposure	SF	0.0278	
24" w/6-1/2" exposure	SF	0.0189	
16" w/4" exposure	SF	0.0313	
16" w/3-3/4" exposure	SF	0.0333	
18" w/5-1/2" exposure	SF	0.0227	
24" w/7-1/2" exposure	SF	0.0164	
24" w/7" exposure	SF	0.0175	
24" w/6-1/2" exposure	SF	0.0189	

QUICK REFERENCE CHART #8
Labor Factors for Sheet Metal

Item Description	Unit	Factor
Flashings		
Aluminum	SF	0.05333
Copper	SF	0.05333
Zinc	SF	0.05333
Fabric backed copper	SF	0.02500
Fabric backed aluminum	SF	0.02500
Mastic backed copper	SF	0.02500
Mastic backed aluminum	SF	0.02286
Fabric	SF	0.02222
Plastic	SF	0.02762
Gutters		
Galvanized steel	LF	0.08000
Aluminum	LF	0.08000
Copper	LF	0.08000
Downspouts		
Aluminum	LF	0.05333
Copper		
3"	LF	0.05333
4"	LF	0.06400
2" x 3"	LF	0.05333
3" x 4"	LF	0.06400
Galvanized steel	LF	0.05333
Louvers		
Aluminum eave vents	LF	0.04571
Attic vent 12" x 24"	EA	0.72727
Foundation vents	EA	0.16000
Brick vents	EA	1.33333
Flashing		
Valley	LF	0.06667
Window & Door	LF	0.06667
Ridge/Hip	LF	0.05714
Gutters	LF	0.08000
Downspout	LF	0.08000
Ridge Roll	LF	0.08000

Chapter 13

SIDING AND EXTERIOR PAINTING

There is a great number of choices in siding, not only in variety but in price. For non-wood residential siding, the wood look is the most popular these days. Modern coatings permit texturing of aluminum to look like wood with rich, long-lasting colors and warranties up to 40 years. Vinyl products, which never need painting, come in a wide variety of wood colors and textures. Vinyl can be molded to duplicate every detail of wood, stone, brick and shake.

New developments in hardboard and plywood offer excellent alternatives to wood boards or shakes. Hardboards are available in prefinished panels. The colors are rustic, and the panels are not easily vandalized, a valuable feature for low-income housing, where many homes need re-siding. Hardboard siding is even available in a stain base, and plywood siding is available in several types of wood, including fir, pine, cedar and redwood.

Finally, there is cedar. It is still the premium product on the market, but because of cedar's cost, it makes the most sense on custom and high-quality remodeling jobs where the money exists to justify its use.

WOOD SIDING

Wood siding comes in a variety of types, sizes and shapes. Types of wood include pines, redwood, cypress, hemlock, spruce, cedar and poplar. Among the classifications of wood siding are bevel, drop and board. There are also shingles and shakes. An advantage to wood is that it is a natural insulator. Before application the wood should be treated with a water repellent, if it has not already been pretreated by the lumber mill. But even if the lumber is pretreated, it is important that cut surfaces should be treated before installation.

Shingles, Shakes and Siding

	per sf
Redwood Bevel Siding	
1/2" x 6" C Redwood	$2.70
1/2" x 8" C Redwood	3.60
1/2" x 10" C Redwood	4.98
Wood Siding	
1" x 6" #1 & 2 Spruce (T&G, rough sawn)	1.68
1" x 8" #1 Pine (Channel Lap)	2.64

Rough Sawn Cedar
7/8" x 12", 8' to 16' (Interior/Exterior, Bd. & Batten) .... 0.74–0.78 ft.
Cedar Shingles and Shakes
Handsplit Shakes (resawn) .... $104.55 sq.
Sidewall Shakes (18" machine-grooved) .... 69.95 sq.
#1 Wood shingles .... 99.00 sq.
Plywood & Aspenite Siding

	Textured (3/8" Grooves, 8" o.c.)			Inverted Batten (1-1/2" Grooves, 12" o.c.)	
Type	**4 x 8-5/8"**	**4 x 9-5/8"**	**Type**	**4 x 8-5/8"**	**4 x 9-5/8"**
Pine	$20.55	--	Pine	$20.95	--
Fir	22.55	25.55	Fir	22.95	26.95
Aspenite	10.95	--	Aspenite	10.95	--

Rough Sawn Fir Plywood (4" x 8" o.c.)
48" x 96" .... $21.40 ea.
Primed Lap Hardboard Siding
12" wide, 10" exp. (80 sq. ft. bundles, 5 pieces 16" long) .... 35.95 bdl
Sheathing (Asphalt impregnated)
4 x 8-1/2" .... 3.59 ea.
4 x 8-25/32 .... 5.19 ea.
Aluminum Siding
White Horizontal (8" x 12' lengths)
    Smooth .... $65.95 sq.*
    Rough Sawn .... 57.95 sq.*
White Vertical (12" x 10' lengths) .... 73.95 sq.*
    *Add $2.00 sq. for color
Soffit (12" x 12') .... 9.95 ea.
Aluminum Siding Accessories
Snap-on Outside
    Corner Post (10') .... $9.95 ea.
    Outside Corner Post (10') .... 9.95 ea.
Individual Corners .... 0.69 ea.
Inside Corners (10') .... 5.79 ea.
Sill Trim (12'-6") .... 3.77 ea.
"J" Channel (1/2", 5/8", 3/4", 1-1/8", 12') .... 2.59 ea.
"L" Channel (12'-6") .... 2.99 ea.
"F" Channel (12') .... 6.49 ea.
Starter Strip (12'-6") .... 2.22 ea.

Crown Moulding (12') .... 4.49 ea.
Western Sill (10') .... 6.89 ea.
Brick Mold Casing (10') .... 6.33 ea.
Backer Plate .... 0.08 ea.
Siding Nails .... 3.59 lb.
Trim Nails .... 5.89 lb.
Casing & Header Cover (12') .... 5.29 ea.
Vinyl Siding
4" White .... 33.00 sq.
3" White .... 44.45 sq.

Bevel and Drop Siding. The labor cost of placing bevel or drop siding varies with the class of work and the method of placement. In the least expensive method, only one end of the siding is squared and the ends of the corners are left rough and covered with metal corner pieces. This method is used extensively in the southern United States, where the long summers and intense heat can cause mitered wood corners to open up.

Where a quality appearance is desired, it is necessary to cut and fit each piece of siding between the window casing and the corner boards. Two masons or carpenters usually work together. It is necessary to square one end of the siding and then measure and cut each board separately to insure a snug fit. This also applies where the siding is mitered at the corners.

Another method of finishing off exterior corners for bevel or drop siding is to install corner boards. These boards are generally made from 1" or 1-1/4" material, depending on the thickness of the siding. They can be plain or molded, depending on the architecture of the house.

Corner boards can be applied to the sheathing and the siding fitted tightly to the narrow edge of the corner board. When this method is used, the joints between the siding and the corner boards should be caulked or treated with water repellent. Corner boards can also be applied over the siding to minimize water infiltrating the ends of the siding.

This latter method is better for paneling. Interior corners are butted against a corner strip consisting of 1" or 1-1/4" material, depending on the thickness of the siding. Bevel siding should have sufficient lap to prevent wind-driven rain from working behind the boards. FHA standards require a minimum 1" lap.

Compare this photo with the one opposite, which shows the same house, but with new windows and siding.

SIDING AND EXTERIOR PAINTING

Quantity of Bevel Siding Required per 100 Sq. Ft. of Wall

Measured Size Inches	Actual Size Inches	Exposed to Weather	Pattern	Add for Lap	B.F. Req'd per 100 S.F. Surface
1/2 x 4	1/2 x 3-1/4	2-3/4	Regular	46%	151
1/2 x 5	1/2 x 4-1/4	3-3/4	Regular	33%	138
1/2 x 6	1/2 x 5-1/4	4-3/4	Regular	26%	131
1/2 x 8	1/2 x 7-1/4	6-3/4	Regular	18%	123
5/8 x 8	5/8 x 7-1/4	6-3/4	Regular	18%	123
3/4 x 8	3/4 x 7-1/4	6-3/4	Rabbetted	18%	123
5/8 x 10	5/8 x 9-1/4	8-3/4	Rabbetted	14%	119
3/4 x 10	3/4 x 9-1/4	8-3/4	Rabbetted	14%	119
3/4 x 12	3/4 x 11-1/4	10-3/4	Rabbetted	12%	117

The above quantities include 5% for end cutting and waste.

Weight of Beveled Siding per 1,000 ft.

Type	Thickness	Lbs. per 1,000 B.F.
Redwood	1/2"	800
Red Cedar	1/2"	600
Redwood	5/8"	1,000
Red Cedar	5/8"	750
Redwood	1"	1,200
Red Cedar	1"	900

Quantity of Drop Siding Required per 100 Sq. Ft. of Wall

Measured Size Inches	Actual Size Inches	Exposed to Weather	Add for Lap	B.F. Req'd per 100 S.F. Surface	Lbs. per 1,000 B.F.
1 x 6	3/4 x 5-1/4	5-1/4	14%	119	1,800
1 x 8	3/4 x 7-1/4	7-1/4	10%	115	2,000

The above quantities include 5% for end cutting and waste.

SIDING AND EXTERIOR PAINTING

Quantity of Shiplap Required per 100 Sq. Ft. of Surface

Measured Size Inches	Actual Size Inches	Add for Lap	B.F. Req'd per 100 S.F. Surface	Lbs. per 1,000 B.F.
1 x 8	3/4 x 7-1/4	10%	115	2,200
1 x 10	3/4 x 9-1/4	8%	113	2,200

The above quantities include 5% for end cutting and waste.

Quantity of Dressed and Matched (D & M) or Tongued and Grooved (T & G) Boards Required per 100 Sq. Ft. of Surface

Measured Size Inches	Actual Size Inches	Add for Width	B.F. Req'd per 100 S.F. Surface	Lbs. per 1,000 B.F.
1 x 6	3/4 x 5-1/4	14%	119	2,100
2 x 6	1-5/8 x 5-1/4	14%	238	2,300

The above quantities include 5% for end cutting and waste.

To prepare exterior stucco walls for new siding, it would probably be necessary to remove the existing stucco. The labor cost per 1,000 sq. ft. is:

	Hours	Rate	Total	Rate	Total
Laborer	1.25	--	--	$21.50	$ 26.88

To add 1/2" plywood sheathing over existing boards, the cost per 100 sq. ft. would be:

	Hours	Rate	Total	Rate	Total
105 sq. ft. sheathing		--	--		$32.00
1 lb. nails		--	--		0.03
Carpenter	1.0	--	--	$25.60	25.60
Helper	0.3	--	--	19.45	5.84
110 s.f. felt		--	--		3.47
Carpenter	0.4	--	--	25.60	10.24
Cost per 100 s.f.			--		$77.18
Cost per s.f.			--		0.77

Installing Plain Bevel Siding. It is particularly important to have the first course level. If the first course is not level and uniform, all subsequent courses will also be off. The position of the first course is determined by measuring down from the bottom of the soffit to 1" below the bottom plate, which will usually be on the foundation. Snap a horizontal chalk line. Do this at 2' intervals across the entire wall. Then, 3/8" furring strips are nailed along the chalk lines to provide support for the starting course.

The butt edge of the first course is laid over the furring strip aligned with the chalk line. Start at one side and work to the other. The first course is nailed to the bottom plate at points just below each stud to mark nailing locations for subsequent courses.

Overlap is measured at the top of the siding to be lapped and a chalk line placed to show the position of the butt edge of the second course. The second course is then started, also at the wall's end, and nailed with a single nail at each stud, tapped flush with the siding face. Siding should fit snug but not tight. Vertical joints should be staggered and positioned so that they fall on studs. Under a window sill, siding should be cut to fit the groove on the bottom of the sill. Over doors and windows, siding should be set on the drip cap.

Labor Placing Bevel Siding

Measured Size Inches	Actual Size Inches	Exposed to Weather Inches	Class of Workmanship	B.F. placed per 8-Hr. Day	Carpenter Hours per 1,000 B.F.
4	3-1/4	2-3/4	Rough Ends	350-400	21.3
4	3-1/4	2-3/4	Fitted Ends	240-285	30.5
4	3-1/4	2-3/4	Mitered Corners	200-240	36.3
5	4-1/4	3-3/4	Rough Ends	415-460	18.2
5	4-1/4	3-3/4	Fitted Ends	285-330	25.0
5	4-1/4	3-3/4	Mitered Corners	240-285	30.5
6	5-1/4	4-3/4	Rough Ends	475-525	16.0
6	5-1/4	4-3/4	Fitted Ends	325-375	23.0
6	5-1/4	4-3/4	Mitered Corners	265-310	28.0
8	7-1/4	6-3/4	Rough Ends	570-620	13.4
8	7-1/4	6-3/4	Fitted Ends	375-415	20.0
8	7-1/4	6-3/4	Mitered Corners	300-350	24.0
10	9-1/4	8-3/4	Rough Ends	650-700	12.0
10	9-1/4	8-3/4	Fitted Ends	440-480	17.5
10	9-1/4	8-3/4	Mitered Corners	375-425	20

SIDING AND EXTERIOR PAINTING

12	11-1/4	10-3/4	Fitted Ends	475-525	16
12	11-1/4	10-3/4	Mitered Corners	400-450	19

Labor Placing Drop Siding

Measured Size Inches	Actual Size Inches	Class of Workmanship	B.F. placed per 8-Hr. Day	Carpenter Hours per 1,000 B.F.
6	5-1/4	Rough Ends	525-575	14.5
6	5-1/4	Fitted Ends	350-400	21.4
6	5-1/4	Mitered Corners	285-325	26.3
8	7-1/4	Rough Ends	600-650	12.8
8	7-1/4	Fitted Ends	415-460	18.2
8	7-1/4	Mitered Corners	325-375	23.0

Estimated Nail Quantities

Size	Length Alum.	Gal.	No. of Lbs. Alum.	Gal.	Lbs. per 1,000 B.F. Alum.	Gal.
6d	1-7/8"	2"	560	190	2	6
7d	2-1/8"	2-1/4"	460	170	2-1/2	6-1/2
8d	2-3/8"	2-1/2"	320	120	4	9
10d	2-7/8"	3"	210	100	5-1/2	11

SHINGLES AND SHAKES

Shingles, like vertical siding and paneling, can be applied over old siding. If the siding is uneven, a 1" x 4" nailing strip is applied horizontally as a nailing base for the shingles. Spacing of the nailing strips depends on the length and exposure of the shingles. Shingles should be applied with rust resistant nails.

Shingles come in 16", 18" and 24" lengths and in random widths from 3" to 14", as well as uniform widths of 4", 5" and 6". The uniform widths are called *dimension shingles* and are applied with tight fitting joints. The random widths are applied with 1/4" to 1/8" spacing between joints to allow for expansion due to changes in weather. Both types can be applied by two basic methods: single-course or double-course.

Single-Course Method. This method consists of simply laying one course over the other, the same as in bevel siding. However, greater exposure is permitted, allowing the use of second-grade shingles, because only half or less of the butt is exposed. The spacing for shingles is determined the same as for bevel siding. A carpenter and helper can apply approximately 500 sq. ft. of single-course wood shingles per 8-hr. day at the following cost per 100 sq. ft.:

	Hours	Rate	Total	Rate	Total
Carpenter	1.6	--	--	$25.60	$40.96
Laborer	1.6	--	--	19.45	31.12
Cost per 100 s.f.			--		$ 72.08

Double-Course Method. This method consists of applying an undercourse of shingles and nailing a second course over it with 1/2" or 3/4" projection of the butt over the lower course. In this method the top course should be first grade, while the undercourse can be a third course or undercourse grade. All joints must be broken so that the vertical butt joints of the top course are 1-1/2" from the joints of the undercourse.

A carpenter and laborer can apply 300 sq. ft. of double-course shingles per 8-hour day at the following cost per 100 sq. ft.:

	Hours	Rate	Total	Rate	Total
Carpenter	2.67	--	--	$25.60	$ 68.35
Laborer	2.67	--	--	19.45	51.93
Cost per 100 s.f.			--		$120.28

Weather Exposure - Shingles
Maximum Exposure

Size	Single Course	Double Course	
		1st Grade	2nd Grade
16"	7-1/2"	12"	10"
18"	8-1/2"	14"	11"
24"	11-1/2"	16"	14"

Shakes. Shakes are somewhat longer than shingles and come in lengths of 18, 24 and 32 inches. Of the three kinds, Handsplit and Resawn are hand-cut and machine finished, while Tapersplit and Straightsplit are mostly hand-cut and finished. Any of the preceding can be applied by the two basic methods of application, single-course or double-course. Nails should be rust-resistant. In the single-course method, it is customary to blind nail the shake by placing the nail 1"

SIDING AND EXTERIOR PAINTING

above the butt line of the next course. Since they are longer, shakes generally have greater weather exposure than shingles.

Weather Exposure - Shakes
Maximum Exposure

Size	Single Course	Double Course Resawn/Taper	Straight
18"	11-1/2", 8-1/2"	14"	16"
24"	11-1/2"	20"	22"
32"	15"		

A carpenter and helper can apply 600 sq. ft. of single-course shakes per 8-hour day at the following cost per 100 sq. ft.:

	Hours	Rate	Total	Rate	Total
Carpenter	1.33	--	--	$25.60	$34.05
Laborer	1.33	--	--	19.45	25.87
Cost per 100 s.f.			--		$59.92

Grades and Sizes of Cedar Shingles

Grades	Sizes	Bundles per Square	Lbs. per Square
1	16", 18" and 24"	4	192
2	"	"	158
3	"	"	144

Estimated Number of Wood Shingles
Required to Cover 100 Sq. Ft.*

Exposure	No. of Shingles Required	Nails Required (lbs.) 3d	4d
4"	990	3-3/4	6-1/2
5"	792	2	5-1/4
6"	660	2-1/2	4-1/4

*Includes 10% waste

Grades and Sizes of Cedar Shakes

Grade	Dimensions	Bundles per Square	Lbs. per Square
No. 1	18" x 1/2" to 3/4"	4	220
Handsplit & Re-sawn	18" x 3/4" to 1-1/4"	5	250
	24" x 1/2" to 3/4"	4	280
	24" x 3/4 to 1-1/4"	5	350
	32" x 3/4" to 1-1/4"	6	450
No. 1 Tapersplit	24" x 1/2" to 5/8"	4	260
No. 1 Straightsplit	18" x 3/8"	5	200
	24" x 3/8"	5	260

Recommended Nails for Wood Shingles

Size	Length	Gauge	Head	Shingles
3d	1-1/4"	14-1/2	7/32"	16" & 18"
4d	1-1/2"	14	7/32"	24"
5d	1-3/4"	14	7/32"	16" & 18"
6d	2"	13	7/32"	24"

3d and 4d nails are used for new construction;
5d and 6d nails are used for re-roofing.

In the double-course method, the second course is nailed 2" above the butt-line. A carpenter and laborer can apply 400 sq. ft. of double-course shakes per 8-hour day at the following cost per 100 sq. ft.:

	Hours	Rate	Total	Rate	Total
Carpenter	2.0	--	--	$25.60	$ 51.20
Laborer	2.0	--	--	19.45	38.90
Cost per 100 sq. ft.			--		$ 90.10

SIDING AND EXTERIOR PAINTING

HARDBOARD LAP AND PANEL SIDING

Hardboard lap siding comes in various sizes. Standard sizes are 8", 10" and 12" wide by 16' long. Hardboard panels come in 4' x 8' and 4' x 9' sizes, but up to 4' x 16' is available if necessary. When stored on the job site, the material should be placed on stringers, avoiding direct contact with the ground and covered with a waterproof covering to prevent direct exposure to the elements.

As with all wood building materials, hardboard siding should never be applied where there is excessive moisture, such as over drying concrete or plaster, or on wet sheathing materials. Where such conditions exist, allow the building to dry before application proceeds.

Vapor barriers of one perm or less (polyethylene film or foil-backed gypsum board, for example) are required on the warm side of the wall in all heated or insulated buildings.

Hardboard siding should never be installed in direct contact with the ground or where water may collect and contact the siding. Allow at least 6" to 8" of space between the siding and the ground or anywhere where water is apt to collect.

The application table below shows the recommended nailing procedure. *Remember: Nails should penetrate a minimum of 1-1/2" into the framing member. Always use corrosion resistant nails.*

Installing Hardboard Lap Siding. Hardboard lap siding is recommended for wood-framed structures. It is not recommended for use as mansard roofing. Stud spacing on exterior walls should be a minimum 16" o.c. When application is direct to studs or over wood board sheathing, a breather-type, not a vapor-barrier, building paper should be used behind the siding. The building paper can be dispensed with when moisture-resistant panel sheathing is used, but it might be required by local code or regulations. A 3/8" x 1-1/2" cant strip is used to start the first course. This is nailed along the bottom of the wall framing as shown in the illustration below.

The bottom of the first course is positioned overlapping the cant strip 1/4" to 1" and nailed, placing nails 16" apart, through the siding and cant strip into the joint, header or sill. Plastic hammer caps are suggested to minimize hammer marks. Nailing should proceed from one end to the other leaving the nail heads slightly above the surface.

Do not set or drive nails especially when siding is installed over soft material such as plastic foam sheathing.

Each course of siding must overlap the previous course by 1". Nails are placed along the bottom edge 16" apart at stud locations about 3/4" above the bottom edge. Vertical joints must be centered over framing members and the siding nailed on each side of the joint. The use of metal joint covers will permit speedier application and offer a more professional appearance and weathertight wall. If joint covers are not used, all joints should be caulked. Siding ends should be spaced 3/16" apart with the joint cover centered, or corner boards may be used.

Where siding is notched out around windows, doors and other openings, shims should be installed behind the siding for continuous support. Nails should be 16" apart through siding and shim. Where siding butts against window, door and other vertical members, there should be a 1/8" space left for expansion. All areas where siding butts around doors, windows and other openings should be caulked. Outside corners can be covered with metal corners, covered or butted against 1-1/8" thick wood corner boards.

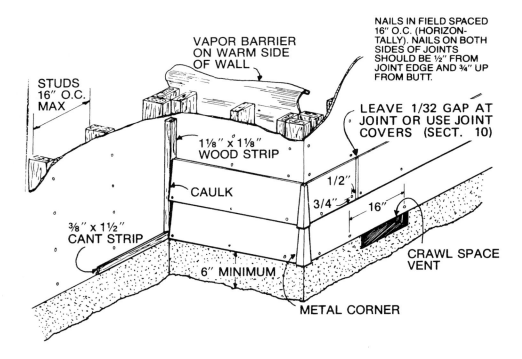

At inside corners, a metal inside corner is installed or the siding is butted against a 1-1/8" sq. piece of wood leaving a 1/8" gap, which is caulked. Good building practices call for keeping siding at least 6" above the finished grade and out of contact with roofs, slabs and places where moisture tends to collect.

SIDING AND EXTERIOR PAINTING

A professional sider can usually apply 450 sq. ft. of hardboard lap siding per 8-hour day at the following cost per 100 sq. ft.:

	Hours	Rate	Total	Rate	Total
Carpenter	1.77	--	--	$25.60	$45.31
Laborer	1.77	--	--	19.45	34.43
Cost per 100 s.f.			--		$79.74

Installing Hardboard Panel Siding. The uses of hardboard panel siding are the same as for hardboard lap siding. The same conditions for moisture apply. Hardboard panel siding should never be applied over wet sheathing.

A vapor barrier, one perm or less, must be installed on the warm side of the wall in all heated or insulated buildings. If plastic foam sheathing is installed on the cold side of the studs, particular attention should be given the vapor barrier because of the low vapor permeability of foam materials. Check the manufacturer's recommendation when applying vapor barriers.

The following application procedure describes installing vertical groove, reverse board and batten and plain hardboard panel siding. When applying vertical grooving or reverse board and batten panels, studs must be 16" o.c. maximum. Plain panels, ungrooved, may be applied over studs 24" o.c. where building codes permit. Unless local codes or building regulations dictate otherwise, building paper may be omitted behind panel sidings having shiplap joints or battens.

Vertical grooved and reverse board and batten panels must be installed with the grooves running vertically. Panel must butt over framing members. The position and spacing of nails is shown in the illustration below. The head should be left slightly above the surface. *It is important not to set or overdrive the nails.*

Square-edge ungrooved panels may be applied over sheathing or direct to studs with the panel edges centered over the studs. Nails are placed 3/8" from panel edges and should penetrate framing at least 1". A gap of 1/16" to 1/8" is recommended between square edge panels, which is covered with a batten strip nailed to the stud.

If siding is to provide racking resistance, nails must be spaced 4" around the perimeter of the panel and 8" at intermediate supports. When applied over structural sheathing, nailing must be spaced 6" at the edges and 12" at intermediate supports.

Panels with shiplap edges can also be applied over sheathing or direct to studs. Shiplaps are available in two widths, 3/8" and 3/4". When nailing, a 3/8" shiplap must be nailed 3/8" from the edge on both sides of the shiplap. For 3/4" shiplap, nailing must be 3/8" from the edge through the underlap and the overlap.

All horizontal joints should have solid wood blocking behind them, and made weathertight. When panels are stacked, as in two- and three-story homes, a metal "z" flashing should be applied to the top of the first course to provide a waterproof joint. Clearance between these panels must be at least 3/8" to 1/2", depending on joist depth and moisture content.

Outside corners are usually finished by applying corner boards as shown in Figure 1. There are several methods for finishing inside corners. These are illustrated in Figures 2, 3 and 4.

A carpenter and helper can usually apply 800 sq. ft. of hardboard panel siding per 8-hr. day at the following labor cost per 100 sq. ft.:

	Hours	Rate	Total	Rate	Total
Carpenter	1.0	--	--	$25.60	$25.60
Helper	1.0	--	--	19.45	19.45
Cost per 100 s.f.			--		45.05

VINYL SIDING - VERTICAL AND HORIZONTAL

Vinyl siding offers a number of advantages over other types of siding. Two of the biggest benefits are long, transferable warranties and the fact that it is maintenance free. In addition to its solid color, it does not peel, blister, flake or rot. It is impervious to termites and other insect pests, requires no painting and does not show scratches. Horizontal vinyl siding comes in two standard styles: *Double 4"* clapboard and *Single 8"* clapboard. Both types come in 12'-6" lengths with 8" exposure. Each comes in a variety of colors suited to all types of homes. When applied with foam polystyrene or fiberboard backerboard, vinyl siding has the added value of thermal and acoustical insulation.

Vertical vinyl siding has the same properties as horizontal. It is used primarily to highlight areas such as doorways, porches, porch ceilings and gable ends. It comes in a *Double 5"* panel, 10' long with 10" exposure. A wide variety of installation accessories are available for both horizontal and vertical vinyl siding, including starter strips, undersill trim, window and door channels, inside and outside corner posts and window and door caps. Warranties up to 40 years are offered. However, in cold climates many carpenters do not recommend vinyl, because it tends to become brittle and when jarred or hit, it will crack and break.

SIDING AND EXTERIOR PAINTING

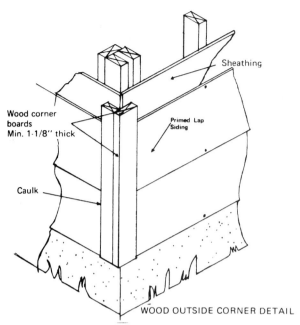

Fig. 1 — WOOD OUTSIDE CORNER DETAIL

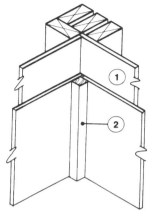

panel siding
1. sheathing or building paper if required.
2. wood inside corner.

Fig. 2

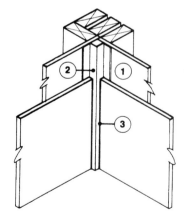

panel siding
1. sheathing or building paper if required.
2. metal inside corner.
3. caulking or sealant.

Fig. 3

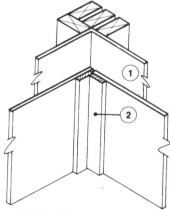

panel siding
1. sheathing or building paper if required.
2. battens.

Fig. 4

Installing Horizontal Siding. For remodeling work it will be necessary to remove shutters, downspouts and outside electrical fixtures. Replace any rotted boards, and nail down any loose ones. If old siding has been removed, apply waterproof sheathing paper. If walls and base are not level and plumb, they should be furred out where necessary. Strapping may be required on uneven outer walls and masonry walls.

Nails should be placed in the center of the slots, not set or flush. Leave a minimum 1/4" clearance at all "J" channels and corner posts to allow for normal expansion and contraction. When temperature is below 32°, leave 3/8" clearance. Nails should be aluminum or some other corrosion resistant metal, 1-1/2" to 2" long, plain shank, 0.12" dia. and 5/16" flat head.

Snap a chalk line for the starter strip. Nail the starter strip every 8", alternating rows of nailing slots. Double nail at the ends. Inside corner posts are positioned 1/4" from top and nailed beginning at the top 12" o.c. Outside corner posts are installed in the same manner.

As installation reaches window and door caps, they should be furred out where necessary for alignment with adjacent panels. In remodeling existing window head flashing is left in place.

"J" channels are installed over window-door caps. A 1/4" tab is cut at each end of the cap and the cap is bent down over the "J" channel on each side to prevent water from running down behind the vertical "J" channel.

After the first panel is locked in the starter strip, backerboard is inserted bevel edge down and toward the wall. Panels are nailed 16" o.c. (not over 8" in areas where severe winds may be encountered). Nails should be centered in nailing slots. Panels should not be pulled up tight, but allowed to hang without strain.

Panels are lapped one-half of the factory notched cutouts. The end of the uncut panel should always lap over the end of the notched panel. Laps should be staggered for best appearance. Undersill trim is fitted under window sills; at the tops of walls against soffits; moldings and to finish off trimmed and fitted cut edges of siding panels anywhere above the level of the starter strip. Caulking should be done whenever required or according to manufacturer's recommendations. When cutting siding to fit around windows and doors, a saw or sharp knife should always be used to cut locking detail and lower edge.

An experienced carpenter and helper should be able to install 500 sq. ft. of horizontal vinyl siding per 8-hour day at the following cost per 100 sq. ft.:

SIDING AND EXTERIOR PAINTING

	Hours	Rate	Total	Rate	Total
Sider	1.6	--	--	$25.60	$40.96
Helper	1.6	--	--	19.45	<u>31.12</u>
Cost per 100 s.f.			--		$72.08

Installing Vertical Siding. Corner posts, window and door channels and window and door caps are installed the same as for horizontal siding. A chalk line is run around the bottom side of the walls and vertical base flashing is installed with nails spaced 8" o.c. Window and door cap is installed above all windows and doors not previously covered. Vertical panels are started from the center of the wall and installed working in both directions. A plumbline is dropped at the center of the wall and a double vertical starter strip nailed 8" o.c. at the top of the slot, 7/8" to one side of the plumbline. Panels are nailed in the center of the nailing slots 16" o.c. An allowance of 3/8" for expansion is left at the top of each panel. Fitting around windows and doors is accomplished the same as for horizontal panels. Gables are finished with 1/2" inverted "J" channel fitted along the gable slope. An experienced sider and helper should be able to install 500 sq. ft. of vertical vinyl siding per 8-hr. day at the following labor cost per 100 sq. ft.:

	Hours	Rate	Total	Rate	Total
Sider	1.6	--	--	$25.60	$40.96
Helper	1.6	--	--	19.45	<u>31.12</u>
Cost per 100 s.f.			--		$72.08

Installing Aluminum Siding. Aluminum siding is manufactured in a variety of durable colors and finishes, including simulated wood. For energy savings it is also available with insulation laminated to the underside. Backerboard made of various insulating materials simply dropped behind the panels is also an option.

Life expectancy of aluminum siding is 20 years or more in most situations. Styles include *Double 4"*, *Double 5"* and *Single 8"* for horizontal installation, board and batten and v-groove in 8", 12" and 16" for vertical installation. Accessories similar to those offered with vinyl siding to facilitate installation and improve finished appearance are available.

Pre-planning and proper preparation of the walls are essential to a smooth-running job that produces quality appearance. Rotten boards should be removed and loose boards should be nailed firmly to framing members. Furring strips should be used to bring up low spots so that nailing surfaces are true and even.

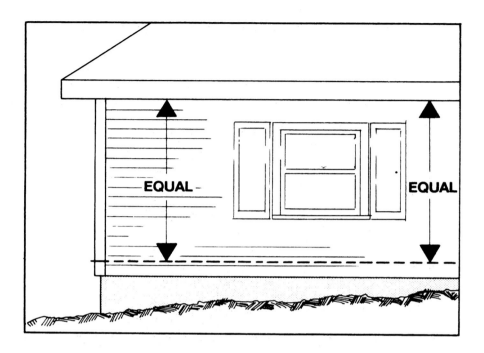

Backer-strips are used with 8" horizontal siding to provide panel support at key points. Outside corner caps are applied one at a time to complete the outside junction of panels. Inside corners are finished similar to those for vinyl.

Aluminum siding is hung on the nails, not nailed to the wall. Nails should not be set or driven flush. In remodeling care should be taken that nails are holding securely and in firm, solid wood. Only aluminum nails should be used, and they should be driven at right angles to the wall, avoiding angling the nail up or down.

Panels should overlap at least 1/2". Except for insulated panels, laps should be supported with backer-strips. Short pieces of siding, 20" or less, should be avoided. Joints or laps should be staggered for strength and appearance. Caulk should be applied around all openings where water has a chance of entering.

An experienced aluminum sider and helper should be able to install 500 sq. ft. of aluminum siding per 8-hour day at the following labor cost per 100 sq. ft.:

	Hours	Rate	Total	Rate	Total
Sider	1.6	--	--	$25.60	$40.96
Helper	1.6	--	--	19.45	31.12
Cost per 100 s.f.			--		$72.08

SIDING AND EXTERIOR PAINTING

Caulking. For caulking around doors and windows, a nozzle 3/8" or larger should be used. Caulking material can be classified into two broad groups. Those in the first group, with a life span of 5 to 20 years, are made from natural and synthetic rubber and are ranked in performance, beginning at the low end with latex, followed by butyl, and then neoprene, which offers the longest life. The second group is the high performance type, which includes polyurethane, polysulfide and silicone. A building caulked ten years ago with one from the first group could already have an infiltration problem today.

One caulker using a cartridge-type gun should be able to caulk about 400 to 450 lin. ft. per 8-hr. day at the following cost per 100 lin. ft.:

	Hours	Rate	Total	Rate	Total
Caulker	1.88	--	--	$19.45	$36.57
Cost per lin. ft.			--		0.37

MASONRY REPAIRS

Cutting Out Old Brick Mortar Joints. Using a carborundum wheel 5" in diameter and 1/4" thick for a cut to a depth of 1/2", the following cost per 100 sq. ft. may be assumed:

	Hours	Rate	Total	Rate	Total
Tuckpointer	2.0	--	--	$26.25	$52.50
Helper	1.0	--	--	20.10	20.10
Cost per 100 s.f.			--		$72.60
Cost per s.f.			--		0.73

Repointing Mortar Old Brick Joints. After the mortar joints have been cut out, two tuckpointers working on the scaffold with one helper on the ground should point from 400 to 500 sq. ft. of brick wall per 8-hr. day at the following labor cost per 100 sq. ft.:

	Hours	Rate	Total	Rate	Total
Tuckpointer	3.5	--	--	$26.25	$91.88
Helper	1.5	--	--	20.10	30.15
Cost per 100 s.f.			--		$122.03
Cost per s.f.			--		1.22

Removing Brick by Hand Chipping. A mason should hand chip around 200 pieces per 8-hr. day at the following labor cost per 100 pieces:

	Hours	Rate	Total	Rate	Total
Mason	4.5	--	--	$26.25	$118.13
Helper	1.0	--	--	20.10	20.10
Cost per 100 pieces.			--		$138.23
Cost per brick			--		1.38

Removing Brick by Pneumatic Pick. Where larger areas of brick are to be broken out, figure two workers at the following cost per 100 sq. ft. of 12" thick wall:

	Hours	Rate	Total	Rate	Total
Labor	18.0	--	--	$21.50	$387.00
Pick Rental		--	--	40.00	40.00
Cost per 100 s.f.			--		$427.00
Cost per s.f.			--		4.27

In replacing masonry veneer, care should be taken to retain the original appearance. After removing all loose mortar and brushing the joint to remove dust and loose particles, the surface should be dampened. The mortar should be tamped well into the joint to achieve a good bond. Pointing should be the same as existing joints.

Care should be taken to keep mortar off the face of the brick or stone. Since older homes frequently used soft bricks and porous stone trim, the entire surface may require treatment with transparent waterproofing. Painted, stained or dirty brick can be cleaned by careful sandblasting, then repainted or waterproofed.

Patching Face Brick. To patch 4" face brick, figure one mason and one helper can lay about 65 sq. ft. per day at the following labor cost:

	Hours	Rate	Total	Rate	Total
Mason	12.0	--	--	$26.25	$315.00
Helper	12.0	--	--	20.10	241.20
Cost per 100 s.f.			--		$556.20
Cost per s.f.			--		5.56

Removing Concrete Block. Block walls can usually be knocked down manually by two laborers. For 12" or 8" block walls, figure 450 sq. ft. per 8-hour day at the following cost per 100 sq. ft.:

SIDING AND EXTERIOR PAINTING

	Hours	Rate	Total	Rate	Total
Laborer	1.78	--	--	$20.10	$35.78
Cost per s.f.			--		0.36

To remove 4" partition block, figure 100 sq. ft. as follows:

	Hours	Rate	Total	Rate	Total
Laborer	4.5	--	--	$20.10	$90.45
Cost per s.f.			--		0.90

Removing Gypsum Tile. Figure one laborer can remove 100 sq. ft. in 2.5 hours as follows:

	Hours	Rate	Total	Rate	Total
Laborer	2.5	--	--	$20.10	$50.25
Cost per s.f.			--		0.50

Patching Building Brick. Where openings are closed up or otherwise patched out in small areas, one mason and one helper should lay about 30 sq. ft. of building brick per day. The labor cost of patching 100 sq. ft. of 12" building brick wall is as follows:

	Hours	Rate	Total	Rate	Total
Mason	27.0	--	--	$26.25	$ 708.75
Helper	27.0	--	--	20.10	542.70
Cost per 100 s.f.			--		$1,251.45
Cost per s.f.			--		12.51

Patching Concrete Block Partitions. One mason and a helper should patch out 150 sq. ft. of 4" block partitioning per 8-hr day. The labor cost of patching 100 sq. ft. of 4" block partition is as follows:

	Hours	Rate	Total	Rate	Total
Mason	5.5	--	--	$26.25	$144.38
Helper	5.5	--	--	16.90	110.55
Cost per 100 s.f.			--		$254.93
Cost per s.f.			--		2.55

Rebuilding Chimneys. On small chimneys one mason with a helper will lay about 500 bricks per day at the following cost per brick:

	Hours	Rate	Total	Rate	Total
Mason	8.0	--	--	$26.25	$210.00
Helper	8.0	--	--	20.10	160.80
Cost per 500 brick			--		$370.80
Cost per brick			--		0.74

SOFFITS

Manufactured soffit systems are convenient and easy to install and offer beauty and long life for the homeowner. They are suitable for roof overhangs, porches, covered walkways and patio overhangs. Systems are aluminum or vinyl, and each comes with a complete line of accessories.

Standard aluminum soffit panels are 12" wide by 12' long. Vinyl panels are approximately 10" wide by 10' long. Both come in solid as well as perforated panels to permit proper breathing. At least 25% of the total installation should be perforated.

Installation. When using the aluminum system, a 40' run of overhang will require 120 running feet of soffit of which 25%, or about 30 lin. ft., should be perforated. The same measure will require 40 lin. ft. of molding or frieze board, 40 lin. ft. of fascia cap and fascia board and 1 lb. of 1-1/2" siding nails and 1 lb. of white trim nails. (If the rafter sidewall is to be covered, figure the area, allowing for some scrap.)

In remodeling, if the existing soffit is level, the rear soffit support can usually be hung so that it coincides with the bottom of the existing soffit. Check the manufacturer's instructions. At appropriate intervals throughout the run, determine the correct length of soffit pieces to be cut.

Slide the first piece of soffit into the rear soffit support and nail through the nailing flange into the bottom of the fascia board. Center nailing is not required in soffit lengths up to 4 lin. ft. The exposed edge of the soffit will be covered by the fascia cover.

Subsequent soffit pieces are locked into place by pulling the piece toward the installer, locking it into the v-groove. Each fourth panel should be perforated. At corners the run can be mounted square, 90° or 45°.

One sider should be able to handle 300 lin. ft. of soffit per 8-hr. day at the following labor cost per 10 lin. ft.:

SIDING AND EXTERIOR PAINTING

	Hours	Rate	Total	Rate	Total
Sider	2.67	--	--	$25.60	$68.35
Cost per 100 l.f.			--		68.35
Cost per s.f.			--		6.84

EXTERIOR FINISH CARPENTRY

Placing Corner Boards and Fascia Boards. When placing wood fascia boards, corner boards and the like, a carpenter should place 175 to 225 lin. ft. per 8-hr. day at the following cost per lin. ft.:

	Hours	Rate	Total	Rate	Total
Carpenter	0.04	--	--	$25.60	$1.02
Cost per l.f.			--		1.02

Placing Exterior Wood Cornices and Verge Boards. When placing exterior wood cornices, verge boards, fascia and the like, consisting of two members, two carpenters working together should place 150 to 175 lin. ft. per 8-hr. day at the following cost per lin. ft.:

	Hours	Rate	Total	Rate	Total
Carpenter	9.82	--	--	$25.60	$251.39
Cost per l.f.			--		2.51

When the exterior is three members (crown mold, bed mold, fascia, etc.), two carpenters working together should place 100 to 125 lin. ft. per 8-hr. day at the following labor cost per lin. ft.:

	Hours	Rate	Total	Rate	Total
Carpenter	14.29	--	--	$25.60	$365.82
Cost per l.f.			--		3.66

If a four-member wood cornice is used, two carpenters working together should place 60 to 75 lin. ft. per 8-hr. day at the following cost per lin. ft.:

	Hours	Rate	Total	Rate	Total
Carpenter	23.53	--	--	$25.60	$602.37
Cost per l.f.			--		6.02

The above quantities and costs do not include blocking out for items such as fascia boards and cornices. An extra allowance should be made for all blocking required.

Placing Brick Moldings. A carpenter should fit and set about 32 lin. ft. of brick molding per hour at the following cost:

	Hours	Rate	Total	Rate	Total
Carpenter	0.03	--	--	$25.60	$0.77
Cost per 100 l.f.			--		76.80
Cost per l.f.			--		0.77

Placing Wood Cupolas. One carpenter should set a prefabricated pine cupola in about two hours:

	Hours	Rate	Total	Rate	Total
Carpenter	2.0	--	--	$25.60	$51.20

Time and material for flashing should be added to the above cost.

EXTERIOR PAINTING

When estimating quantities of paint and other exterior finishes, the actual surface to be painted should be measured as accurately as possible, either by taking measurements from the plans or directly from the building. From these measurements it should be possible to estimate the quantity of materials required, but labor costs present a much more difficult problem, owing to the different classes of work and the difficulties encountered. For instance, the covering capacities of paint will be the same for a plain surface or a cornice, but the labor cost of painting the cornice is considerably more because of the height of the cornice above the ground and the amount of cutting or trimming necessary. Care must be used when pricing any piece of work, because conditions on each job are different.

Clapboard or Drop Siding Walls. Obtain the actual area of all walls and gables. Add 10% to allow for painting the bottom edge of the siding boards. Do not deduct for openings that are less than 10'-0" x 10'-0".

Shingle Siding. Obtain actual area of all walls and gables and multiply by 1.5. Do not deduct for openings less than 10'-0" x 10'-0".

SIDING AND EXTERIOR PAINTING

Eaves. For plain eaves to be painted the same color as the side walls, obtain total area of eaves and multiply by 1.5. If eaves are to be painted a different color from side walls, obtain area and multiply by 2.

Eaves with rafters running through, obtain area and multiply by 3.

Eaves over brick, stucco or stone walls, obtain area and multiply by 3.

For eaves over 30'-0" above ground, multiply by 1.5 for each additional 10'-0" in height.

Cornices, Exterior. For plain cornices obtain the total area and multiply by 2. For fancy cornices, such as cornices containing dentils, obtain the total area and multiply by 3.

Downspouts and Gutters. Determine the area of plain downspouts and gutters and multiply by 2. For fancy downspouts and gutters, find the area and multiply by 3.

Blinds and Shutters. Figure the area of one side of plain blinds or shutters. Multiply by 2. For slatted blinds or shutters, measure the area of one side and multiply by 4.

Columns and Pilasters. For plain columns or pilasters, determine the square feet of surface area; if fluted, obtain the area and multiply by 1.5; if paneled, obtain the area and multiply by 2.

Lattice Work. Measure one side and multiply by 2, if painted one side only; if painted two sides, obtain the area and multiply by 4.

Porch Rail, Balustrade, Balusters. If the balustrade is solid, add one foot to the height to cover the top and bottom rails. Multiply the length by the height. If painted two sides, multiply by 2.

If the balustrade consists of individual balusters and handrail, multiply length by height and then multiply the result by 4. For handrail only, if under 1'-0" girth, figure as 1'-0" by length and multiply by 2.

Moldings. When cut in on both sides, figure 1 sq. ft. per lin. ft., if under 12" girth. If over 12" girth, take actual measurement.

Doors and Frames, Exterior. It costs almost as much to paint a small door as a large one. Do not figure any door as less than 3'-0" x 7'-0". To allow for the door frame, add 2'-0" to the width and 1'-0" to the height. For instance, a 3'-0" x 7'-0" door would be figured as 5'-0" x 8'-0" or 40 sq. ft.

If it is a sash door, containing small lights of glass, add 2 sq. ft. for each light: a 4-lt. door would be figured with an additional 8 sq. ft.; a 12-lt. door would be an additional 24 sq. ft.; etc. If painted on both sides, obtain the sq. ft. area of one side and multiply by 2. For door frames only, where no door is hung, remember to figure the area of both sides of the frame into the total area to be covered.

Windows, Exterior. It costs almost as much to paint a small window as a large one. Do not figure any window as less than 3'-0" x 6'-0". Add 2'-0" to both the width and height of the opening to take care of the sides and head of the frame and the outside casing or brick mold, and multiply to obtain the area. For example, to a window opening that is 3'-0" x 6'-0", add 2'-0" to both the width and height, making 5'-0" x 8'-0" or 40 sq. ft. of surface.

If sash contain more than one light each, such as for casement sash, add 2 sq. ft. for each additional light. A 6-lt. window would contain an additional 12 sq. ft.; a 12-ft. window, 24 sq. ft. additional.

Roofs. For flat or nearly flat roofs, figure the actual area. For roofs with a 1/4 pitch, add 25%. For roofs with a 1/3 pitch, 33.3%. For roofs with a 1/2 pitch, add 50%.

Fences. For plain fences, measure one side and multiply by 2; for picket fences, measure one side and multiply by 4.

Covering Capacity of Oil Base Paint on New Exterior Wood. The number of square feet that a gallon of oil base paint will cover depends a great deal on the surface to be painted: kind of wood, degree of roughness, etc. Some woods are more porous than others and absorb more paint. Much depends also on how the paint is brushed out. Some painters brush the paint out more and cover more surface.

The priming coat, properly applied, should cover 450 to 500 sq. ft. per gallon. The second coat should cover 500 to 550 sq. ft. per gallon, and the third coat should cover 575 to 625 sq. ft. per gallon.

For exterior trim painting, with measurements taken in accordance with standard methods, the priming coat should cover 750 to 850 sq. ft. per gallon; the second coat should cover 800 to 900 sq. ft. per gallon and the third coat should cover 900 to 1,000 sq. ft. per gallon.

Bear in mind the above coverages are based on surfaced lumber. The covering capacity will be greatly reduced when applied to rough boards.

Painting Old Exterior Wood Surfaces. When estimating repaint work on old exterior wood surfaces, if the existing surface is sound and in good condition, two coats should be sufficient. Paint coverages should be about the same as for the second and third coats on similar new work.

If the old surface shows cracking, blistering, scaling and peeling, the old paint should be removed by using either a paste or liquid paint remover of the slow drying type, or a blow torch and scraper.

If the old paint is removed completely, three coats of paint should be applied the same as recommended for new work.

SIDING AND EXTERIOR PAINTING

Painting Wood Shingle Siding. Wood shingle siding should receive three coats of paint on new work. Paint coverages per gallon should be as follows: first coat, 300 to 325 sq. ft.; second coat, 400 to 450 sq. ft.; third coat, 500 to 550 sq. ft.

Wood shingles that have been previously painted with an oil paint and are in suitable condition for repainting should receive two coats of paint. The first coat should cover 400 to 450 sq. ft. per gal. and the second coat 500 to 550 sq. ft. per gal.

Painting Exterior Wood Floors and Steps. When painting exterior wood floors and steps with oil base paint, material coverages per gallon should be as follows: first coat, 325 to 375 sq. ft.; second coat, 475 to 525 sq. ft.; third coat, 525 to 575 sq. ft.

Painting Brick, Stone, Stucco and Concrete. When painting masonry surfaces that are reasonably smooth, one gallon of oil base paint should cover 170 to 200 sq. ft. for the priming coat; 350 to 400 sq. ft. for the second coat and 375 to 425 sq. ft. for the third coat.

Painting Metal Work. The area that any paint may be expected to cover on metal work will vary with the surface to be painted. Badly pitted or rough metal will require more paint than a perfectly smooth surface. The covering capacity will also vary with the temperature, consistency of the paint and the effort behind the brush. For ordinary smooth surfaces, one gallon of paint should cover 550 to 650 sq. ft., one coat.

Aluminum Paint. Aluminum paint is used for both priming and finishing coats on metal surfaces. The covering capacity varies with the condition of the surface, but on smooth surfaces one gallon should cover 600 to 700 sq. ft. with one coat.

COVERING CAPACITY OF PAINTS

Several factors influence covering capacity. Brushing the paint out thinner covers more surface. Dark paint hides the surface better than light paint and so can be brushed out thinner than the lighter colors. Also a rough surface will require considerably more paint than a smooth surface. Soft porous wood will absorb more oil and require more paint than close grained lumber. Another consideration is paint ingredients. Different brands and types of paint vary in covering capacity or hiding power.

Shingle Stain. When staining wood shingles (after it has been laid), one gallon of good stain should cover 120 to 150 sq. ft. of surface for the first coat. The second and third coats will go farther, because the wood does not absorb as much stain as on the first coat and should cover 200 to 225 sq. ft. per gallon. One gallon of good stain should cover about 70 sq. ft. of roof with two coats or 50 sq. ft. with three coats.

If the shingles are dipped in stain before laying, it will require 3 to 3-1/2 gallons of stain per 1,000 shingles. When dipping shingles in stain, only two-thirds of the shingle need be dipped. It is unnecessary to dip the end of the shingle that is not exposed.

Covering Capacity of Oil and Spirit Stains. When staining finishing lumber, such as birch, mahogany, oak and gum, one gallon of oil stain should cover 700 to 725 sq. ft. One gallon of spirit stain should cover 500 to 600 sq. ft.

Covering Capacity of Oil Base Paint on New Exterior Wood. A great deal depends on the surface to be painted; that is, factors such as the kind of wood and the degree of roughness. Some woods are more porous than others and absorb more paint. Much depends on the way the paint is brushed out, because some painters brush the paint out more covering more surface with less paint.

The priming coat, properly applied, should cover 450 to 500 sq. ft. per gallon. The second coat should cover 500 to 550 sq. ft. per gallon, and the third coat should cover 575 to 625 sq. ft. per gallon.

For exterior trim painting, the priming coat should cover 750 to 850 sq. ft. per gallon; the second coat should cover 800 to 900 sq. ft. per gallon and the third coat should cover 900 to 1,000 sq. ft. per gallon.

Bear in mind that the above coverages are based on surfaced lumber. The covering capacity will be greatly reduced when applied to rough boards. The percentage for overhead and profit should be added to the costs listed on the following pages.

Cost of Burning Off Paint from 100 Sq. Ft. of Old Siding

	Hours	Rate	Total	Rate	Total
Painter	2.85	--	--	$24.50	$69.83

Cost of Sanding and Puttying 100 Sq. Ft. Plain Wood Siding and Trim

	Hours	Rate	Total	Rate	Total
Painter	0.5	--	--	$24.50	$12.25

SIDING AND EXTERIOR PAINTING

Cost of Sanding and Puttying 100 Sq. Ft. Outside Trim Only

	Hours	Rate	Total	Rate	Total
Painter	0.86	--	--	$24.50	$21.07

Cost of 100 Sq. Ft. (1 Sq.) Two-Coat House Paint Applied to Plain Wood Siding and Trim

	Hours	Rate	Total	Rate	Total
Priming Coat					
0.24 gal. paint		--	--	$17.75	$ 4.26
Painter	0.95	--	--	24.50	23.28
Second Coat					
0.19 gal. paint		--	--	17.75	3.37
Painter	0.85	--	--	24.50	20.83
Cost per 100 sq. ft.	1.80		--		$51.74

Cost of 100 Sq. Ft. (1 Sq.) Three-Coat House Paint Applied to Plain Wood Siding and Trim

	Hours	Rate	Total	Rate	Total
Priming Coat					
0.24 gal. paint		--	--	$17.75	$ 4.26
Painter	0.95	--	--	24.50	23.28
Second Coat					
0.19 gal. paint		--	--	17.75	3.27
Painter	0.85	--	--	24.50	20.83
Third Coat					
0.17 gal. paint		--	--	17.75	3.02
Painter	0.75	--	--	24.50	18.38
Cost per 100 sq. ft.	2.55		--		$73.04

Painting costs vary according to quality, color and material prices. Add for surface preparation.

Cost of 100 Sq. Ft. (1 Sq.) Two-Coat House Paint Applied to Exterior Trim Only

	Hours	Rate	Total	Rate	Total
Priming Coat					
0.13 gal. paint		--	--	$18.77	$2.44
Painter	1.40	--	--	24.50	34.30
Second Coat					
0.12 gal. paint		--	--	18.77	2.25
Painter	1.12	--	--	24.50	27.44
Cost per 100 sq. ft.	2.52		--		$66.43

Cost of 100 Sq. Ft. (1 Sq.) Three-Coat House Paint Applied to Exterior Trim Only

	Hours	Rate	Total	Rate	Total
Priming Coat					
0.13 gal. paint		--	--	$18.77	$2.44
Painter	1.40	--	--	24.50	34.30
Second Coat					
0.12 gal. paint		--	--	18.77	2.25
Painter	1.12	--	--	24.50	27.44
Third Coat					
0.11 gal. paint		--	--	18.77	2.06
Painter	1.00	--	--	24.50	24.50
Cost per 100 sq. ft.	3.52		--		$92.99

Cost of Applying One-Coat House Paint to One Average Wood Half-Screen (Screen Wire Not Painted)

	Hours	Rate	Total	Rate	Total
0.02 gal. paint		--	--	$17.75	$0.36
Painter	0.14	--	--	24.50	3.43
Cost per screen			--		$3.79

SIDING AND EXTERIOR PAINTING

Cost of Applying One-Coat House Paint to One Average Wood 2-Lt Storm Sash

	Hours	Rate	Total	Rate	Total
0.04 gal. paint		--	--	$17.75	$0.71
Painter	0.29	--	--	24.50	7.11
Cost per storm sash			--		$7.82

Paint costs vary according to quality used, color and local material prices. Add for surface preparation.

Cost of 100 Sq. Ft. (1 Sq.) Two-Coat House Paint Applied to Old Wood Surfaces in Good Condition

	Hours	Rate	Total	Rate	Total
First Coat					
0.19 gal. paint		--	--	$17.75	$ 3.27
Painter	1.0	--	--	24.50	24.50
Second Coat					
0.17 gal. paint		--	--	17.75	3.02
Painter	0.8	--	--	24.50	19.60
Cost per 100 s.f.			--		$50.39

Cost of 100 Sq. Ft. (1 Sq.) One-Coat Oil Paint Applied to Brick or Smooth Concrete Surfaces

	Hours	Rate	Total	Rate	Total
0.54 gal. paint		--	--	18.77	$10.14
Painter	0.93	--	--	24.50	22.79
Cost per 100 s.f.			--		$32.93

Cost of 100 Sq. Ft. (1 Sq.) One-Coat Oil Paint Applied to Rough Plaster or Stucco Surfaces

	Hours	Rate	Total	Rate	Total
0.67 gal. paint		--	--	17.75	$11.89
Painter	1.1	--	--	24.50	26.95
Cost per 100 s.f.			--		$38.84

Labor and Materials Required for Various Painting Operations

Exterior Work—Residential Description of Work	Painter No. Sq. Ft. per Hr.	Painter Hours 100 Sq. Ft.	Material Coverage Sq. Ft. per Gal.
Sanding and Puttying Plain Siding and Trim	200-210	0.50	
Sanding and Puttying Outside Trim Only	115-120	0.86	
Burning Paint off Plain Surfaces	40-50	2.20	
Burning Paint off Wood Siding	30-40	2.85	
Exterior Brush Painting Plain Siding And Trim......Priming Coat	100-110	0.95	400-450
Second Coat	115-125	0.85	500-550
Third Coat	125-135	0.75	575-625
Exterior House Painting Rubberized Woodbond......One Coat	125-135	0.75	450-500
Exterior Trim Only......Priming Coat	65-75	1.40	750-850
Second Coat	85-95	1.12	800-900
Third Coat	95-105	1.00	900-1,000
Oil Paint—Shingle Siding*......First Coat	115-125	0.82	250-300
Second Coat	150-160	0.65	375-425
Oil Paint—Shingle Roofs*......First Coat	100-120	0.90	120-150
Second Coat	155-175	0.60	220-250
Stain—Shingle Siding*......First Coat	75-85	1.20	120-150
Second Coat	115-130	0.82	200-225
Stain—Shingle Roofs*......First Coat	145-155	0.67	100-120
Second Coat	190-210	0.50	170-200
*If surfaces are exceptionally dry, increase labor hours and decrease covering capacity.			
Asbestos Wall Shingles......First Coat	60-70	1.50	150-180
Second Coat	85-95	1.11	350-400
Brick Walls, Oil Paint......First Coat	100-120	0.93	170-200
Second Coat	140-160	0.67	350-400
Third Coat	155-175	0.60	375-425
Porch Floors and Steps, Oil Paint......First Coat	235-245	0.42	325-375
Second Coat	270-280	0.36	475-525
Third Coat	285-295	0.35	525-575

SIDING AND EXTERIOR PAINTING

Labor and Materials Required for Various Painting Operations—Con't.

Exterior Work—Residential

Description of Work	Painter No. Sq. Ft. per Hr.	Painter Hours 100 Sq. Ft.	Material Coverage Sq. Ft. per Gal.
Waterproof Cement Paint—Smooth Face Brick............First Coat	175-185	0.55	90-110
Second Coat	260-270	0.38	140-160
Clear Waterproof Paint—Smooth Face Brick............First Coat	200-210	0.50	500-510
Second Coat	200-210	0.50	590-610
Stucco, Medium Texture, Oil Paint............First Coat	90-100	1.10	140-160
Second Coat	150-160	0.65	340-360
Third Coat	150-160	0.65	340-360
Stucco, Medium Texture, Wpf. Cement Paint............First Coat	130-140	0.74	90-110
Second Coat	190-210	0.50	125-145
Exterior Masonry, Stucco, Asbestos Cement Board and Shingle Siding— Rubberized Flat Finish............First Coat	100-120	0.91	325-375
Second Coat	165-175	0.59	375-425
Concrete Walls, Smooth, Wpf. Cement Paint............First Coat	170-180	0.57	110-130
Second Coat	275-285	0.36	150-170
Concrete Floors and Steps—Floor Enamel............First Coat	250-280	0.38	440-460
Second Coat	190-210	0.50	575-625
Third Coat	200-220	0.48	575-625
Cement Floors—Color Stain and Finish............First Coat	350-370	0.28	475-525
Second Coat	280-300	0.35	450-500
Fences, Plain, Average............First Coat	125-135	0.77	450-470
Second Coat	190-200	0.51	530-550
Fences, Picket, Average............First Coat	140-150	0.70	630-650
Second Coat	160-170	0.60	650-675
Fences, Wire-Metal, Average............First Coat	100-110	0.95	900-1,000
Second Coat	140-150	0.70	975-1,125
Shutters, Average, Each Coat............No. of Shutters	2-3	0.40	11-13
Downspouts and Gutters, Paint............First Coat	170-180	0.57	540-560
Second Coat	185-195	0.53	575-600
Screens, Wood Only, Average, Each Coat............No. of Screens	6-8	0.14	45-55
Storm Sash, 2 Light, Average, Each Coat............No. of Sash	3-4	0.29	23-27

Spray Painting—Residential Work

Description of Work		Painter No. Sq. Ft. per Hr.	Painter Hours 100 Sq. Ft.	Material Coverage Sq. Ft. per Gal.
Brick, Tile and Cement—Cement Water Paint	First Coat	265-275	0.37	90-110
	Second Coat	325-350	0.30	140-160
Brick, Tile and Cement—Oil Paint	First Coat	240-260	0.40	250-275
	Second Coat	240-260	0.40	400-450
Brick, Tile and Cement—Synthetic Resin Bound Exterior Paint	First Coat	275-290	0.36	125-150
	Second Coat	335-360	0.30	175-200
Rough Brick, Tile, Cement and Stucco*—Plastic Paint	One Coat	265-275	0.37	65-75
Asbestos Wall Shingles*	First Coat	100-110	1.00	150-175
	Second Coat	150-175	0.60	225-250
Stucco*—Exterior Cement Water Paint or Synthetic Resin Bound Exterior Paint	First Coat	200-225	0.47	90-100
	Second Coat	265-285	0.37	125-150
Stucco*—Oil Paint	First Coat	160-175	0.60	200-225
	Second Coat	225-240	0.43	350-400
Shingle Roofs—Oil Paint	First Coat	260-275	0.37	125-150
	Second Coat	275-300	0.35	200-225
Shingle Roofs—Stain	First Coat	275-300	0.35	125-150
	Second Coat	160-175	0.60	200-225
Shingle Siding*—Oil Paint	First Coat	175-200	0.55	125-150
	Second Coat			225-250
Shingle Siding*—Stain	First Coat	175-200	0.55	125-150
	Second Coat	190-215	0.50	200-225

*Trim Must be Figured Separately as Hand Work

Chapter 14

WINDOWS AND DOORS

In selecting windows and doors, energy efficiency is as important as aesthetic considerations. Manufacturers have continually improved designs over the last several years to make their products even more energy efficient, and there is a wide selection of energy-efficient designs to fit traditional and contemporary decor.

Energy loss through windows happens in three ways: infiltration, conduction and radiation. Radiation is not as serious as infiltration and conduction, because it works in two directions. In winter there can be a desirable heat gain from radiation that exceeds the window's conduction loss. Where radiation is a problem, reflective film or tinted glass are good solutions. A window that opens and closes can also help solve the problem.

Frequently, infiltration and conduction can be corrected simply by installing storm windows or weather-stripping, but if the infiltration is really bad, replacement is the best answer.

Adding another layer of glass will reduce the conduction loss by 50%. A storm window added to an existing window with insulating glass will reduce conduction by 30%. If the existing window must be replaced and there are no storm windows, thermalized replacement windows with double or triple insulating glass are an excellent solution.

Window styles vary geographically. In the West horizontal sliding windows are favored. In the Northeast, North Central and South, builders favor casement and double hung windows. Single hung windows are mainly found in the South.

In window specifying the width is given first, followed by the height and then the number of pieces of glass (called lights). Finally, the window style is indicated. For example: *28-1/2" x 24", 2 Lts., DH* means the window glass is 28-1/2" wide, 24" high, there are two pieces of glass and the window is a double hung unit.

COMMON WINDOW STYLES

Window style and size should take into account several things. Although in most cases window height is a standard 6'-8", the same as door height, sill height will vary with the style and size of the window. A rule-of-thumb is 1'-0" in living and family rooms, 2'-6" for dining rooms and 3'-6" for kitchens. Exterior viewing should be convenient whether a person is standing or sitting.

How many windows should a rooms have? As a general rule the glass area of a room should be no less than 10% of the floor area, and at least 4% of the same floor area should be devoted to windows that open and close for ventilation. There should also be a balance between fixed and operating windows.

Window styles are varied and versatile. Component parts can be quickly disassembled for easy cleaning, and there is an infinite variety of configurations for opening and closing. Really, the only limit is how much money the homeowner is willing to spend. But when it comes to replacing old windows, most homeowners want, if not the best, a window of considerably higher quality than what they presently have. It is usually wise to recommend the best quality window and work down.

Windows and doors are a neglected portion of the remodeling market. Siding jobs are the perfect opportunity to suggest a new entry door and windows. Not only do you increase your profits, you serve your clients well by suggesting a way to upgrade a home's appearance, which is what they are mainly interested in. Changing old windows for new ones can give an old home a new look, and by careful selection it can give it a new architectural style without extensive changes.

The trend is to use stock or *standard* windows. *Custom* windows are still made, but cost makes them impractical, and anyway, most standards come with a variety of options. Some manufacturers make what is called a *standard special*. These windows are generally designed for a particular marketing area or housing development. For example, say that a tract builder developed an area containing 500 homes 15 to 20 years ago. Recognizing the need for replacement windows in these developments, makers may have developed special stock windows designed for the particular development. Often they are not listed in the catalog. If a contractor is bidding a job in a large, older development and windows are involved, a query to the original window supplier might show that they have a special stock window for that particular development.

Thanks to the versatility of standards, as well as the variety of styles, most stock windows can be made to fit openings in older buildings through the use of accessories offered by many manufacturers. Frame extenders of different widths adapt stock windows to a wide range of wall thicknesses. Exterior vinyl and metal casings not only simplify installation but help seal a home against energy loss. Sill nose covers, jamb clips and metal drip caps all contribute to the ease of installation and help make stock windows energy efficient and adaptable to most existing window openings.

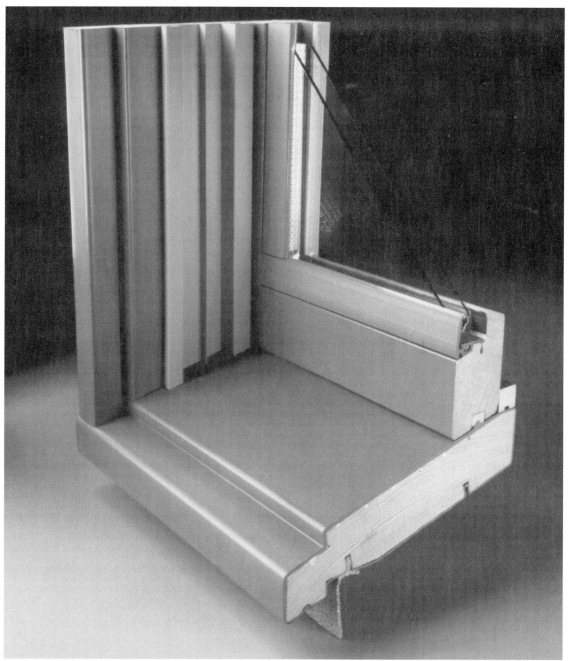
Double Hung Window Unit, PVC Sheath with Double Pane Insulated Glass

WINDOW TYPES

Double Hung Windows. The double hung window (vertical sliding) is still among the most familiar and popular. It has two sashes that slide up and down vertically. The sashes are provided with springs, balances or compression weather stripping that help hold it in place in any location. A number of manufacturers offer windows in which the sash can be easily removed for cleaning or painting. They can be grouped in a number of ways, double with mullion in-between, triple with a fixed window in the middle or even up to five to create a window wall. Many come with snap-in wood muntins to create a Colonial or Cape Cod effect. Patterns include horizontal bars, divided lights or diamond lights.

Double hung windows do have limitations. Only half of the window can be open at one time for ventilation. If the sashes are not removable, they are difficult to clean. Located over a sink or counter, they can be hard to open.

Casement Windows. Casement windows (side-hinged) swing out horizontally, generally with a crank. With a storm window or insulated glass, they are extremely energy efficient. Screens are mounted on the inside. The swinging windows are on the outside and swing out. They are generally factory assembled and come with all hardware in place. Their biggest advantage is that the entire window can open for ventilation. Another is that they can be cleaned from the inside, unless the hinges are set so as to prevent access to the outside glass. A casement window by design will be tighter than a double hung window.

Horizontal Sliding. In horizontal sliding windows, the sash moves sideways in metal or plastic tracks. These windows can also be grouped. If the sash is removable, they are easy to clean, but like the double hung, only half of the window can be used for ventilation, and they are not always easy to clean.

Awning Windows and Hopper Windows. If a window is top-hinged, it is called an awning window. If it is bottom-hinged, it is a hopper window. Both can have one sash operable and one fixed, or both operable. The awning window swings out, while the hopper window swings in. Both have an upward airflow. Open outward, the awning window directs air upward. However, if open inward, it will direct the air downward. The hopper window, which only opens inward, directs the airflow upward. The awning type is operated with a push-bar or crank, while the hopper generally has a simple lock handle at the top. Both windows can be grouped and easily cleaned from the inside unless hinges interfere. Bottom-hinged windows are generally located low in the wall and are not considered viewing windows.

WINDOWS AND DOORS

Jalousie Windows. A jalousie window consists of a series of horizontal clear or opaque glass slats that usually open outward with a crank. Because of their low insulating value, in colder climates they require a full storm window. They have good ventilation characteristics, and cleaning is easy from the inside.

Fixed Windows. Fixed windows do not open, so they do not require screens or hardware, but neither do they provide ventilation. They can be 6' or 8' across, excellent for viewing and providing lots of light. Cleaning, however, necessitates a trip outside. They are frequently used in combination with a double-hung window on each side. Usually they consist of single light of insulated glass or double glazed in a wood sash. They can, however, be installed without sash by simply setting the glass into rabbeted frame members and held in place with stops.

Glass blocks are another type of fixed window, commonly used in bathrooms and basements. They come in both light-diffusing and light-directing styles. All are 3-7/8" thick and have face sizes of 6" x 6", 8" x 8" and 12" x 12". Blocks add privacy and cut off unpleasant views, reduce disturbing noises and can accent decorative schemes. Regular masonry tools are used for installation.

REMODELING USING STANDARD WINDOWS

In replacing windows, all the basic rules of good construction apply. Usually, one dimension will be "on the money." If it isn't, there are many accessories to make standard windows fit odd-size openings. Frame expanders and expander receptors make it possible to adapt standard windows to fit different wall thicknesses. Metal jamb fastening clips and rigid vinyl auxiliary casing can also be used to expedite installation. Odd-shaped openings can be blocked with 2" x 4"s. Nailing trimmer studs at one or both sides will frequently shorten the width of a rough opening to size.

Windows are not a simple consideration. Remodeling contractors who want a first-class job should contact a window manufacturer, many of which employ architects or architectural representatives. Not only will they assist in selecting type and style, they can help with any structural or installation problems.

WINDOW REMOVAL

Removing Sash and Frame. *Single, Double Hung Window.* Frame removal might not be necessary. Removing only the sash will reduce the cost of replacement. All that is required is to remove the stops, cut the sash cords and remove the parting strip and cord pulleys. Wire brush the opening and remove any storm window hardware. Allow 30 minutes per window. If the complete frame is removed, add another 30 minutes.

Closing Old Window Openings. Place 2" x 4" studs 16" on center, aligning them with existing studs under the window. Toenail studs to the header and sill using 8d or 10d nails. Install sheathing and insulation and apply vapor barrier to inside face (warm side) of framing being sure to overlap any existing vapor barrier. Figure one worker can enclose 20 sq. ft. of wall space per hour at the following cost per sq. ft.:

	Hours	Rate	Total	Rate	Total
Carpenter	1.0	--	--	$25.60	$25.60
Cost per s.f.			--		1.28

INSTALLING REPLACEMENT WINDOWS

Without Reframing. Where blocking is required, the blocking is secured to the framing using 16d nails. Be sure blocking is level and plumb. Standard window installation clips are used to set the window through opening. Insulate all sides. Install new stops and interior trim. Caulk and install exterior trim. Allow 1.5 hours for one carpenter to complete the preceding. If opening and closing hardware are required, add 15 minutes per window to the preceding.

Raising Sill Height. In older windows sometimes the jamb dimensions are such that the width of a standard replacement window will fit, but the height is too short. The opening can be resized by removing the old apron, stool and sill. Measure down from the top of the window opening equal to the new window height and install a rough sill and cripple studs to the required height. A new subsill is installed on the rough sill and the replacement window installed using standard installation technique.

Allow one carpenter about 2 hours to complete such an installation.

Complete Reframing. Window suppliers generally have tables from various manufacturers showing individual glass, sash and rough opening size. However, window sizes vary among manufacturers. When the manufacturer is known, it is best to follow its recommendations. Otherwise, additional blocking or shimming might be required.

If the manufacturer is not known, rough openings can be figured as follows. For a single, double hung window, add 6" to the total glass width for the rough opening width and 10" to the total glass height for rough opening height. For double sash casement windows, add 11-1/4" to the total glass width and 6-3/8" to the total glass height.

Be sure to install new insulation all around replacement windows.

For combination windows figure the total width of the multiple opening by adding the width of the individual sash opening to the width of the mullions plus 2" for the overall rough opening. However, a window manufacturer or distributor should be consulted before any window replacement is bid regardless of the kind of opening to be filled.

Where the desired opening is larger than the existing opening, complete reframing is generally necessary. If the wall is load-bearing, the opening will require temporary support and a new header (3'-6" or more). To eliminate inside patching on one side of the new opening, one side should be positioned 1-1/2" from an existing stud. Corners should be avoided and enough room allowed for trim. The position for new studs can be determined by adding the frame width of the new window plus 3-1/2".

The new header is installed, supported by a jack stud nailed to the existing studs. Header length should be the new window frame plus width dimension plus 3-1/2". The opening is finished off to height size by installing a new rough sill on cripple studs as described previously. One carpenter should be able to reframe 20 sq. ft. per hour at the following cost per sq. ft.:

	Hours	Rate	Total	Rate	Total
Carpenter	1.0	--	--	$25.60	$25.60
Cost per s.f.			--		1.28

Although header size will vary, the following can be used for estimating:

Span (Feet)	Header
3-1/2	2, 2" x 6"
5	2, 2" x 8"
6-1/2	2, 2" x 10"
8	2, 2" x 12"

Most replacement windows are factory assembled and ready for installation. The general procedure is the same regardless of style or manufacturer. However, it is important that the manufacturer's procedure be checked for any specific instruction. For example, one might suggest removing the sash from the frame. Another may recommend that sash, frame and bracing be installed as a unit before cross-bracing is removed. When siding is applied over sheathing, the window is installed first and then the siding reapplied.

If the siding is horizontal, strips of 15-lb. asphalt felt should be slid between sheathing and siding around a window opening. If the casing does not have

WINDOWS AND DOORS

built-in caulking, a bead of caulking should be placed over the siding behind the head and side casing.

Generally, the frame and sash are inserted as a unit, squared and shimmed, if necessary, then nailed in place with 10d nails. Double hung windows should be checked for operation while nailing to be sure the sash works freely.

It is customary when installing a window over a panel siding to run a bead of caulk at the junction of the siding and sill and install quarterround over the caulking. Figure one carpenter can complete the above installation in 2 hours. However, if the window is on the second story, double the installation time. For larger windows or difficult locations, figure it will take two carpenters 4 hours.

Labor to Install Windows
Average Conditions

	Hrs.
Replace windows and frame existing opening:	
Frame building	4
Masonry building	6
Replace sash only (per pair)	2
For one sash	2
Picture window	12
Install mullion window and frame opening	8
Frame new opening:	
Single	4
Mullion	6
Windows with liners, masonry frame, add	2
Trim interior, any window, add	2

EXTERIOR DOORS

Doors are as guilty as windows when it comes to energy loss, particularly through infiltration. There is a variety of energy efficient doors to choose from. Wood and metal, usually steel or aluminum, are standard, but there are also fiberglass composite and vinyl doors. All of these are available in several styles, designs and patterns.

When it comes to insulation efficiency, steel clad doors seem to have the edge, with R-values in the 15 plus range not uncommon. These insulation ratings are achieved by using polyurethane foam and honeycomb cells as the core.

Other features include adjustable thresholds, various types of factory installed weather-stripping and strike plates that can withstand impacts of up to 500 lbs., lessening the chance of breaking and entering. Steel adapter frames that fit into existing frames also add to the security. Special thresholds also prevent frost buildup. Thermal breaks or barriers divide the outside from the inside of the door, eliminating the possibility of any frost or condensation on the inside face.

Factory finish helps keep maintenance costs down, and being metal, all are guaranteed to never warp, crack, shrink or swell. Metal doors also have excellent fire ratings, up to 1-1/2 hour B.

With a door jamb made of composite material and a steel entry door, this popular type of door system is typical of contemporary doors that are made for energy efficiency and maximum security.

Although most wood doors do not have the high R-values of insulated metal, advances in wood technology have given them new stability, and problems of distortion are not as great as they once were.

WINDOWS AND DOORS

Flush Doors. Wood flush doors are made with plywood or other wood type facing applied over a frame. The core can be hollow or solid. Solid cores are generally woodblock or particle board. Solid core doors are preferred for exterior doors, because their construction minimizes warping and distortion.

Metal flush doors are generally made from steel facing with a core that is filled with polyurethane and some type of honeycomb for insulation.

Panel Doors. Panel doors also come in wood and metal. Wood panel doors have solid vertical members called stiles, solid cross members called rails and thin strips that divide the stiles and rails into panels. In addition to the flush style, metal doors also come in embossed designs. Both types come with glass panels.

Glazed Doors. Glazed doors are usually French doors. The rails and stiles are divided into lights by muntins. They are frequently used as patio, porch or terrace doors. They can be hung in a variety of ways, such as singly or in pairs. Snap-in muntins can be added to create a traditional appearance and provide ease of maintenance.

Exterior doors are 1-3/4" thick and 6'-8" high. Wood, steel and aluminum are all popular. The recommended size for entrance or front doors is 3'-0" wide, while side and rear doors can be from 2'-8" to 3'-0" wide. Jambs can be wood or metal. Jamb widths vary according to the material and the kind of wall into which they will be set. Wood jambs are usually 5-1/4" for lath and plaster and 4-1/2" for drywall. However, they can be easily cut down or expanded to fit most openings. Stock metal jambs are manufactured in a variety of widths and recommended for specific applications. Standard widths are made for lath and plaster, concrete block, brick veneer and drywall. These are 4-3/4", 5-3/4", 6-3/4" and 8-3/4". Jambs for drywall, however, are sized differently. Common widths of metal jambs available for this application are 5-1/2" and 5-5/8".

Although most sills are made of oak, when softer woods are used, the assembly should include a metal nosing and wear plate.

Removing Old Doors. A clean rough opening is mandatory if a new door is to function properly. After removing the old door from its hinges, remove inside casing and outside brickmold. Side jambs are generally removed by cutting in half and prying away from studs. If nails are too long, either cut them with a hacksaw or drive them flush and sink them. Pry off the head jamb and sill. Check the subfloor to be sure it is level and flush. If not, it must be filled in. Plumb the opening to be sure it will accept the new unit. Figure one carpenter can accomplish the work at 1.5 hours per door. For cleanup and removal, add 20 minutes.

INSTALLING NEW DOORS

Relocating Doors. It may be desirable to move a door to a new location, which involves closing the old opening and creating a new one. Openings are closed by installing 2" x 4" vertical studs (assuming the wall is a 2" x 4" stud wall) spaced no more than 16" apart, preferably placing them in line with existing studs. Toenail new studs to old door header and to the floor using three 8d or 10d nails at top and bottom. Next, cover the opening area with sheathing that is the same or similar material and thickness as the existing sheathing. Add insulation and apply a vapor barrier over the inside face of the framing (warm side of the wall.) Be sure the vapor barrier covers the rough framing and overlaps any existing barrier already installed. Figure one carpenter can complete the preceding in 4 hours at the following cost per opening:

	Hours	Rate	Total	Rate	Total
Carpenter	4.0	--	--	$25.60	$102.40

Remember that you will need to account for finishing off the wall. This work is not included in the figure above.

Cutting New Opening. If an opening is new, or an existing opening is not square, it will probably have to be rough framed with a header and *cripple* or *door buck* studs. The rough opening should be approximately 2" to 2-1/4" higher and 2-1/2" to 3" wider than the door to be installed. For the header double 2" x 6"s are nailed to the studs on each side of the opening. Door bucks are nailed on each side of the opening with 12d nails, spaced about 16" apart and staggered. The sill should rest on the floor framing, which in most cases must be cut out to a depth that will place the sill even with top of the finished floor. After the sheathing and paneling have been placed over the framing, leaving only the rough opening, the door frame is ready for installation. A carpenter should be able to rough frame a standard 3'-0" x 6'-8" door including the necessary cutouts in 3 hours at the following cost per door opening:

	Hours	Rate	Total	Rate	Total
Carpenter	3.0	--	--	$25.60	$76.80

Installing Door Frame. Apply a bead of caulk on both sides and on top of the cutout or install 15-lb. asphalt felt. Tip the frame into place and brace, using wood shingles as wedges. Square up until true and plumb. Place nails 3/4" in from outer edge, 16" on center, and set with a nail set. Nail casing to the front

WINDOWS AND DOORS

edge of jamb with 10d casing nails spaced 16" on center. Figure one carpenter can install a conventional exterior frame and casing with oak sill in about 2 hours.

	Hours	Rate	Total	Rate	Total
Carpenter	2.0	--	--	$25.60	$51.20

Hanging the Door. After determining the hang of the door, mark both door edges and corresponding jambs accordingly. To operate easily, the door should conform to the shape of the finished opening, less 1/16" clearance at the sides and top. If there is no threshold, bottom clearance should be 1/16". With threshold, there should be 1/8" clearance above the threshold. Check the jambs for trueness and transfer any irregularities to the door. Any trimming to reduce the door width should be done on the hinge side. Bevel the lock edge so that it will clear the door when it is closed. Exterior doors, unless otherwise specified, are hung with three hinges. The top of the top hinge is installed 7" from the top of the door. The bottom of the bottom hinge is installed 11" from the bottom of the door. The center hinge is centered between the top hinge and the bottom hinge. Both the door and jambs are mortised for setting the hinges. The hinges most commonly used are 3-1/2" or 4" loose-pin butt mortise hinges. Since most doors come already cut out for hardware, all that remains is to install the lockset and striker on the door jamb. Figure one carpenter can size the door and install the hinges and lockset, including strike plate, in about 1.5 hours:

	Hours	Rate	Total	Rate	Total
Carpenter	1.5	--	--	$25.60	$38.40

Weather-stripping. Today all doors and windows should be weather-stripped regardless of climate. This includes new units as well as old. Some metal clad doors feature a magnetic weather-strip that seals much like a refrigerator. The most accepted procedure is to caulk around the frame or jamb using latex caulking compound and applying a 1/4" bead. Figure a carpenter can caulk 200 lin. ft. per hour at the following cost per lin. ft.:

	Hours	Rate	Total	Rate	Total
Carpenter	1.0	--	--	$25.60	$25.60
Cost per l.f.			--		0.13

Storm Doors. Energy costs being what they are, so-called "storm doors" are as practical in warm climates as they are in cold. A combination storm and screen door requires about 1.5 hours to install.

	Hours	Rate	Total	Rate	Total
Carpenter	1.5	--	--	$25.60	$38.40

Sliding Glass Doors. There are many types of sliding glass doors, including metal, wood and metal clad, which is wood covered with metal, with a factory applied finish. In addition to there being a style to fit almost any type of architecture, today's sliding glass doors are specifically designed to save energy.

Most are manufactured in frame sizes to fit existing openings and consequently are ideal for remodeling. Furthermore, they are specifically designed to replace thermally inefficient aluminum sliding glass doors. Options such as double glazing and tinted glass provide additional energy savings. They also come with one, two and three side lights.

If space allows for hinged swinging doors, a new popular replacement for sliding doors is an atrium or concept door. These doors fit the same opening, but one panel is hinged rather than sliding. The seal is much tighter, and to some extent, they also provide better security.

Removing Old Doors. Since most new frames are sized to take advantage of the existing rough opening, it is best to remove the entire frame. After lifting out the screen, remove any existing trim on the outside. The operating panel is moved to the middle of the frame, raised and pulled out of the track. Stationary panels are detached by removing the holding plates, then moved to the center and lifted out in the same manner as the operating panel. The frame is usually held in place with screws or by a holding flange underneath the siding. Sometimes a power saw is required to cut the flange after the frame has been removed. Clean the rough opening and check for plumb, level and square. Check rough opening dimensions. Figure one worker can complete this work in 2 hours.

	Hours	Rate	Total	Rate	Total
Carpenter	2.0	--	--	$25.60	$51.20

If the existing rough sill requires replacement in order to provide continuous sill support for the new door, add 1 hour to the above.

WINDOWS AND DOORS

Labor to Install Doors
Average Conditions

	Days
Inside door and jamb	1/2
Exterior door and jamb	1/2
Trim door	1/4
Cutout for window:	
Masonry wall	1-1/2
Frame wall	1/2
Perfatape wall, interior	1/4
Pocket doors (pair)	1/2
Replace interior door in existing frame	1/2

Electronic door openers take a little over 2 hours to install:

	Hours	Rate	Total	Rate	Total
Electrician	2.25	--	--	$29.65	$66.71

Material Prices

Jambs and Frames (A-Grade) White Pine

Interior jambs (32" headers) — Each
- Clear/drywall (for staining) $27.12
- Clear/plaster (for staining) 28.75
- "B"/drywall (for painting) 15.92
- "B"/plaster (for painting) 19.12

Exterior Frames

- With oak sill (36" header) 60.72
- Without sill (36" header) 43.12
- Oak sills
  - 32" 16.88
  - 36" 17.52
- Sliding Door 75.12

Flush Doors

(1-3/8" Hollow Core) Size

	Lauan	Birch
24" x 80"	$29.15	$36.08
26" x 80"	30.75	37.68
28" x 80"	30.75	37.68
30" x 80"	30.75	37.68
32" x 80"	32.88	39.92
36" x 80"	35.12	42.48

Exterior Size (1-3/4")

	Lauan		Birch	
	Hollow	Solid	Hollow	Solid
30" x 80"	$35.12	59.12	46.32	71.92
32" x 80"	38.32	60.72	47.92	73.52
36" x 80"	38.32	62.32	51.12	76.72

Pre-Hung Doors (1-3/8" hollow-core Lauan)

Includes one-side casing (17L) and stop and hardware.

24" x 80"	$82.48
28" x 80"	84.72
30" x 80"	84.72
32" x 80"	86.32
36" x 80"	87.92

Add: Ex. side of casing, $11.18; keyed lockset, $9.60.

Locksets and Hinges

	Style A	Style B	Style C
Passage	$10.22	$17.95	$17.95
Privacy	11.98	21.85	21.85
Entry	23.28	29.68	29.68
Hinges	3.02 pr.		

WINDOWS AND DOORS

Bi-fold Doors

Lauan flush doors (2-panel)
        24" x 80-1/2"........$50.48
        30" x 80-1/2"..........54.32
        36" x 80-1/2"..........62.32

Lauan flush doors (4 panel)
        48" x 80-1/2"..........83.12
        60" x 80-1/2"..........86.32
        72" x 80-1/2"..........92.72

INTERIOR DOORS

Interior doors generally come in two styles, flush and panel. There are also sliding, bifold and accordion-fold door. Flush and panel doors are usually 1-3/8" thick, while folding and sliding doors are usually 1-1/8" thick. Accordion folding doors range in thickness based on weight, size and material that varies from cloth to wood. Flush doors are usually hollow core, consisting of a light frame covered with plywood or hardboard. Finishes come in various woods, such as oak, birch or mahogany, which are generally finished naturally. Hardboard and non-selected grades are usually painted.

Panel doors have solid stiles, rails and panels of various types. The folding louvered door is generally used for closets because of its ventilation characteristics. Folding and sliding doors are also used for wardrobes and pantries.

Pocket doors or doors that slide into the wall are used where space to swing is limited, or where they would block a hallway.

Door widths can be subject to building regulations, so it may be advisable to check to see if any exist before a door is installed. Minimum width is usually 2'-6" for bedrooms and other rooms; 2'-4" for bathrooms and 2'-0" for small closets or linen closets.

Standard interior door height is 6'-8" for first floor doors, while doors of 6'-6" are frequently used on second floors, basements or in homes that are a story-and-a-half.

Hanging Interior Doors. For new or added doors, the opening should be rough framed similar to exterior doors. Hinged doors should open and swing in the direction of entry, preferably toward a blank wall. They should not be obstructed by other swinging doors or swing into a hallway.

Figure 2-1/2" plus the door width for horizontal and 2" plus the door height for vertical framing. Where wood casing is used, the head and side jambs are the same widths as the overall wall thickness. Where metal casing is used with drywall, the jamb width is the same as the stud width.

Jambs can be purchased in precut sets or complete with stops and the door hung in the frame. The prehung door is the simplest and most economical to install. Usually, the stops are temporarily nailed in place until the door is hung. These are usually 7/16" thick and can be from 3/4" to 2-1/2" wide. They are mitered and installed at the junction of the head and side jambs. A "sanitary stop" consisting of a 45° bevel cut at the bottom of the stop 1-1/2" above the floor will make cleaning or refinishing of the floor easier.

When installing the new door frame, the hinge side of the frame is fastened first. Shingle wedges are used between the side jamb and the rough door buck to plumb the jamb. Additional wedges are placed at the hinge and latch locations and along the top. The jamb is nailed using two 8d nails at each wedge, and after installation, the wedges are trimmed flush with the wall.

The door is fitted to the frame using the following clearances: 1/8" for the knob-side and top, 1/16" on the hinge side and 1/2" or more at the bottom, particularly if it is to swing over carpeting. The hinges should be the proper size for the door they will support. For 1-3/8" doors, the best size is 3-1/2" x 3-1/2". Two hinges are sufficient for most hollow-core doors. However, if the door is heavier, three are better.

Locks come in three styles: entry lock sets, which are keyed and decorative; bath or bedroom lock sets, called *privacy lock sets,* which have an inside lock control and a safety slot for opening the door from the outside; and passage or latch set, which has no locking device.

Lock sets are usually purchased with the door and may or may not be already installed. If not, the manufacturer provides directions that should be carefully followed. Lock sets should be installed so that the doorknob is 36" to 38" above the floor.

The stops, which were temporarily nailed, can now be permanently installed. Casing is positioned about 3/16" from the face of the jamb and nailed, spacing nails about 16" apart.

Metal casing can be nailed to the door buck and the drywall inserted and then nailed to the studs. Or the casing can be fitted to the drywall and the drywall attached by nailing through the casing and the drywall into the stud.

Generally, one carpenter can complete the above procedure, including framing and installing the door, hinges and hardware in 1.5 hours at the following labor cost per door:

WINDOWS AND DOORS

	Hours	Rate	Total	Rate	Total
Carpenter	1.5	--	--	$25.60	$38.40

Bypass Sliding Doors. Bypass sliding doors can be used to enclose a variety of storage areas. Their advantage is no floor space is lost to door swing. Installation is usually in a standard door frame, to which an upper track is hung and concealed behind a piece of trim mounted below the head jamb. The rough opening is framed in the same way as for a conventional swinging door. The doors, which contain rollers, are guided on the floor by a small guide track. The rollers are adjustable so that the door can be aligned and plumbed with the opening. Standard sizes are 1-3/8" thick, 6'-8" or 7'-0" high and any width. Be sure to consult the installation instructions, because rough opening sizes vary among manufacturers. One carpenter can usually hang a bypass door 6'-8" x 6', including hardware, at the following cost per installation:

	Hours	Rate	Total	Rate	Total
Carpenter	3.0	--	--	$25.60	$76.80

Bifold Doors. Bifold doors come in a variety of widths and heights. Their advantage is that when they are open, the entire storage area is accessible. Wood and metal are the types most commonly used. Bifold doors can be installed in regular door frames, and openings are framed the same as for a swinging door. Bifold doors operate on upper and lower tracks. A single carpenter can install a four-panel bifold door, including installing the tracks, in about 2.3 hours at the following cost per installation:

	Hours	Rate	Total	Rate	Total
Carpenter	2.3	--	--	$25.60	$58.88

Folding Doors. Folding doors come in a variety of materials and prices. Top of the line are the wood folding doors, offered in a choice of veneers, such as oak, birch, walnut or mahogany. They are available finished or sanded and ready to stain or paint. Frequently, they are used to close off laundry or storage areas. The better quality wood doors, however, can be used to divide rooms and when closed, look like paneling. In addition to convenience, an outstanding feature is that they fold inside their own doorway, taking little room and allowing complete access to the area they enclose. Most come assembled with hardware.

Two-panel units enclose areas up to 26-1/2" wide, while four-panel units will enclose openings up to 72" wide. For larger openings or folding wall room dividers, widths up to 40'-3" are available. Standard finished heights are 6'-8" 7'-6" and 8'-0".

Installation is reasonably simple. One carpenter can generally hang a double folding door in a four-panel unit in 2 hours at the following cost per installation.

	Hours	Rate	Total	Rate	Total
Carpenter	2.0	--	--	$25.60	$51.20

Labor to Set Finish Hardware

Type	Hours per Each
Rim Lock	0.5
Mortised Lock	1.0
Cylinder Lock	0.5
Front Ent. Cylinder Lock	2.0
Surface Door Closer	1.0
Concealed Door Closer	3.0
Sash Lift and Lock	0.5
Kickplate	1.0

Chapter 15

PORCHES AND DECKS

Before patios and decks, the porch was the center of summertime outdoor activity. The word "porch" evokes the image of a large front porch on a frame house. There is a swing seat where a family's young women can meet gentlemen callers and where the inhabitants can escape the heat of a stuffy house on a summer night. It evokes a time before the invention of air-conditioning, when houses were open at night to the sounds of the neighborhood. People escaped the heat by leaving their houses at night, and community activity provided a source of entertainment to be enjoyed from the front porch.

This description may sound a bit idealized, but it is a fairly accurate picture of life in many towns and cities before air conditioning and suburban sprawl evolved into a different kind of outdoor living, one where privacy and intimacy was important. The patio and deck would seem more suited to shutting out the neighborhood, rather than letting ones home life spill outdoors.

Porches are certainly still around and quite popular. Many older homes and urban two-flats and three-flats have porches front and rear. However, decks and patios appear to go far beyond the function of a porch, and much more than the latter, they are used as an additional room, where inhabitants engage in any number of activities—eating, sleeping, reading, entertaining—you name it.

PORCHES

The labor cost for porch work is highly variable because of the vast differences in style and construction of porches and the amount of detail involved. In this section we have offered several approaches to assist the reader in estimating porch remodeling and repairs. And don't forget the cost of permit fees.

Labor to Wreck Porch
(Average)

1 Story	1 day
2 Story	3 days
3 Story	4 days

Replacement
Rear Open Porches - Labor Only
(under average conditions)
Size: 7' x 20'

1 Story	8 days
2 Story	12 days
3 Story	16 days

Porch Repairs - Labor Only

6" x 6" column (not over 12')	4 hrs.
6" x 6" lookout	4 hrs.

Stairs

Front (not over 7 risers, 5' wide)	8 hrs.
Rear (per run, not over 8 risers)	8 hrs.
Winders: Add 50%	
Oversize: Add 25%	

Joists

Single Face (not over 16')	2 hrs.
Single Regular (not over 16')	2 hrs.
Complete Replacement (7' x 12')	4 hrs.

Rails (per lin. ft.)

Picket	10 mins.
Three member	6 mins.
Front Chippendale with 1/3 balustrade	15 mins.

Materials
(Approximate Cost Midwest)

6" x 6" Posts (12' lengths)	$38.50
Lookouts (9' lengths)	27.50
Boxed Columns (12' length)	33.00
Joists (2" x 8")	0.77 lin. ft.
Fir Flooring	1.65 sq. ft.
Railings	
Rear Picket	3.30 ea.
Rear 3-member	2.75 ea.
Front (molded top, 1-3/8") balustrade	6.60 ea.

Typical Porch Replacement Costs
(Materials and Labor)

Stoops. Size: 4' x 6'. No roof, average seven steps including rails and pickets. Total: $583.00. For larger sizes, figure $16.50 per sq. ft. additional.

Rear Open Porch - Type I
(see illustration below)

Size: 7' x 22'
Steps
1" x 4" T&G Flooring
2" x 8" Joists and Rafters
Roof Deck: 5/8" plywood
Roll Roofing
6" x 6" Columns and Lookouts
Picket Railings

1 Story $2332.00
2 Story 3712.50
3 Story 5434.00
Note: Add 6% for each foot in width over 22 feet. Smaller widths, no credit.

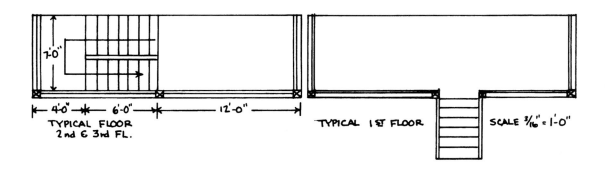

Add for:
Winders .... $11.00 each
Hot Roof......................... 165.00
Piers................................ 55.00 each

Debris Removal (minimum):
1 Story $77.00
2 Story 110.00
3 Story 143.00

Rear Open Porch - Type II

Size: 7' x 22'
Materials, add-ons and
debris removal same as
for Type I.

1 Story $2332.00
2 Story 4603.50
3 Story 6380.00

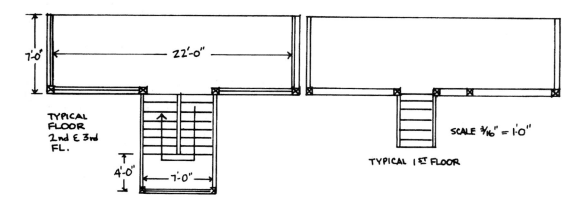

Placing Plain Porch Columns. When placing plain square or turned porch columns, such as are commonly used for rear porches and other inexpensive porches, a carpenter should place one post in about 0.75 hours at the following labor cost:

	Hours	Rate	Total	Rate	Total
Carpenter	0.75	--	--	$25.60	$19.20

Placing Porch Top and Bottom Rail and Balusters. When placing the type of wood top and bottom rail and wood balusters generally found on front porches, a carpenter should complete 15 to 20 lin. ft. of rail per 8-hour day at the following cost per lin. ft.:

	Hours	Rate	Total	Rate	Total
Carpenter	0.46	--	--	$25.60	$11.78

PORCHES AND DECKS

When placing top and bottom rails with open balusters or using matched and beaded ceiling, such as is often used for less expensive grades of work, a carpenter should complete 35 to 45 lin. ft. of rail per 8-hour day at the following cost per lin. ft.:

	Hours	Rate	Total	Rate	Total
Carpenter	0.2	--	--	$25.60	$5.12

Framing and Erecting Exterior Wood Stairs for Rear Porches. When framing and erecting outside wood stairs for rear porches, such as are found on two-flat and three-flat apartment buildings, the stringers are 2" x 10" or 2" x 12" stock, with treads and risers nailed on the face of the stringers. It requires 18 to 22 hours carpenter time per flight of stairs. This is for ordinary stairs having 14 to 18 risers extending from story to story. The labor cost per flight should be as follows:

	Hours	Rate	Total	Rate	Total
Carpenter	20.0	--	--	$25.60	$512.00

If the stair consists of two short flights with an intermediate landing between stories, it will require 12 to 13 carpenter hours per flight, or 24 to 26 hours per story, including platform, at the following labor cost:

	Hours	Rate	Total	Rate	Total
Carpenter	25.0	--	--	$25.60	$640.00

Framing and Erecting Wood Stairs with Winders. Where wood stairs to rear porches have 4 to 6 winders in each story, figure 24 to 28 hours carpenter time at the following labor cost per story:

	Hours	Rate	Total	Rate	Total
Carpenter	26.0	--	--	$25.60	$665.60

Clear Ponderosa Pine Porch Material

Kind of Molding	Size in Inches
Square Baluster Stock	1-1/8" x 1-1/8"
Square Baluster Stock	1-3/8" x 1-3/8"
Square Baluster Stock	1-5/8" x 1-5/8"
Balustrade Cap	3-5/8" x 1-5/8"
Balustrade Shoe	3-5/8" x 1-5/8"
Rabbeted Porch Jamb	1-1/2" x 3-1/2"
Plowed Porch Shoe	1-1/2" x 3-1/2"
Top Rail	3" x 2-1/4"
Top Rail	3-3/4" x 1-3/4"
Bottom Rail	3-1/2" x 1-3/4"

WOOD PATIOS AND DECKS

The size and structure of a deck is limited only by the imagination—and the budget. It can be a simple square or an elaborate, multi-level structure. Pools and hot tubs can be surrounded by the deck, or pleasant garden areas included. Seats that double as railings, tables that double as seats, stairs leading off to another deck level are all parts of a well designed plan.

Patios should harmonize both with nature and with the existing structure. Mature trees provide welcome shade, a hillside offers a dramatic view, or a door makes the deck an extension of a dining room or bedroom. When the structure faces south or west, an awning or cover may be needed as protection from the sun. Lattice work or walls might be used to hide an unwanted view. The possibilities are unlimited, making it easy to overcome virtually any design problem.

For example, the land near the house is uneven, and a paved patio is not practical. The deck can be built near ground level with short posts to overcome the uneven ground. A low-level deck can also be built to cover a concrete patio that has cracked or become unsightly.

In preparing a desk estimate, be sure to consider three basics that affect the total time needed to build the deck:

(1) *Location:* Is the deck near the front of the house, or in the back? Because it is often located at the back of the house, allow adequate time to move all building materials from their delivery point to the area where you will be working.

(2) *Level:* Will the floor be at ground level or elevated to an upper level of the house? The floor level will influence the size of lumber needed for beams and joists, length of the posts and the time necessary for moving material.

(3) *Structure:* How elaborate will the finished structure be? Multiple levels, unusual shapes and sizes or other unusual features add to the time required for planning, cutting and placing the material during construction. Making a detailed diagram is wise, because it will aid in planning for material quantities and in building the deck.

Planning should also include careful consideration of the types and sizes of material to be used for posts, beams and joists, and the decking. The following guidelines cover each of the major steps in building an average 16' x 16' deck at the second story level.

Posts. Foundation posts for decks are usually 4" x 4" lumber or metal pipe. To avoid sinking the posts are set in concrete to an adequate depth that freezing will not affect them (6" below frost line). Or where the ground is solid, they are set on concrete pads or pre-formed concrete anchors.

Figure about one hour per post to dig the hole, set in concrete and brace it to assure that it is plumb. A power auger is often be needed.

Beams and Joists. Steel beams can be used, but 2" x 6" lumber is common for beams up to 16' in length. Many carpenters prefer 2" x 6" joists set on 24" centers. Centers can be up to 36", but 16" centers can be used for extra strength.

Allow approximately one hour per beam (three 16' beams would be need for a 16' x 16' deck) including placing the beam, leveling and bolting in place. For joists on the deck described above, placing all joists 24" on center, figure about 5 hours, including placing the necessary fascia.

Decking. Typical material choices for decking include construction grade lumber, cedar, redwood or specially pressure-treated lumber for outdoor use. Today, there are also any number of engineered wood products, too. The 2" x 6" lumber is preferred for strength, but 1" x 6" can be used, if joists are 16" on center. Galvanized nails should be used to prevent corrosion.

Allow 8 hours to cut and nail all the decking lumber and trimming the ends. If joists are placed on longer or shorter centers, more cutting time may be necessary.

Railing. To cut and place railing on three sides of the deck, allow about 8 hours, including bolts where needed.

Stairs. An upper level deck usually includes a set of outdoor stairs. Figure 8 hours to cut and place one full flight with a railing.

Door. If the deck is added outside a room where no door exists, allow 8 hours to frame and install a standard door. Sliding glass doors will require 12 to 16 hours, depending on how much reinforcing must be done.

The estimate should include approximately 8 hours of time for a helper who will be needed to place heavy items and to move material. For the typical 16' x 16' deck with all items as listed, labor is as follows:

	Hours	Rate	Total	Rate	Total
Carpenter	49.0	--	--	$25.60	$1254.40
Helper	8.0	--	--	19.45	155.60
Cost per s.f.			--		5.55

When the deck is constructed at ground level, growth of vegetation beneath the deck should be prevented. A simple and effective method is to fill the area with pea gravel over a heavy plastic sheet to prevent growth of grass and weeds.

Deck and Patio Lumber Prices
Pressure treated, can be stained, painted or left natural.

Size	Price per lin. ft.	Size	Price per lin. ft.
2" x 4" (8' x 10')	$0.39	2" x 10" (12' x 16')	$1.22
2" x 4" (12' x 14')	0.44	2" x 12" (12' x 16')	1.63
2" x 4" (16')	0.46	4" x 6" (8'-14')	1.22
2" x 6" (8' x 14')	0.65	4" x 6" (16')	1.28
2" x 6" (10',12',16')	0.69	4" x 6" (18' x 20')	1.42
2" x 6" (18', 20')	0.73	4" x 6" (22' x 24')	1.68
2" x 6" (T & G)	0.72	6" x 6" (8'-14')	1.82
2" x 8" (8' x 10')	0.83	6" x 6" (16')	2.02
2" x 8" (12'-16')	0.90	6" x 6" (18' x 20')	2.15
4" x 4" (8')	0.82	6" x 6" (22' x 24')	2.54
4" x 4" (10' x 16')	0.90	1" x 6" Fencing	0.35

Deck and Patio Accessories

Step Brackets (7" rise, 10" tread)	$7.79 ea.
Step Brackets (6" rise, 16" tread)	10.13 ea.
Rail Post Bracket	4.41 ea.
Metal Post Bracket	7.14 ea.

PORCHES AND DECKS

Translucent Roofing (White or Green)
- 26" x 8' corrugated .... 9.53 ea.
- 26" x 10' corrugated .... 12.94 ea.
- 26" x 12' corrugated .... 15.54 ea.

Porch Flooring (1 x 4 P/L)

T&G, kiln-dried, sanded .... $0.26 ft. (600 M)

Ornamental Iron Railing

4' railing	$14.24 ea.	Floor flanges	$3.17 ea.
5' railing	18.14 ea.	Lamp tongue	3.17 ea.
6' railing	21.52 ea.	Flat column	15.30 ea.
Newel Post	6.23 ea.	Corner column	36.34 ea.
Adj. Fittings	4.15 ea.		

Vapor Barrier (.004)

6 x 100	$20.74 roll	20 x 100	$48.04 roll
8 x 100	24.64 roll	20 x 100 (blk)	49.34 roll
10 x 100	25.94 roll	10 x 25	10.65 roll
12 x 100	32.44 roll	15 x 25	15.54 roll
16 x 100	40.24 roll	20 x 25	19.44 roll

Chapter 16

CITY COST INDEXES

The wage rates used in the cost tables found throughout this book are what we have determined to be national averages as of 1998. These prevailing U.S. rates come from surveys of collective bargaining agreements. They do not take into account merit shop wage rates, and they may not fit your business or locality. For this reason we strong encourage users to develop their own unit prices, based on the man-hour productivity rates shown on the tables, using their own average wage rates.

For those who wish to use the unit prices that we give, there will still be some need of adjustments for your area. The following table gives labor cost adjustment factors for 97 U.S. cities. Multiply an amount shown on a cost table by the factor for your city.

City	Cost Factor
Akron, OH	0.995
Albany, NY	0.869
Albuquerque, NM	0.756
Anchorage, AK	1.404
Atlanta, GA	0.751
Baltimore, MD	0.863
Billings, MT	0.987
Birmingham, AL	0.633
Boise, ID	0.989
Boston, MA	1.299
Buffalo, NY	1.055
Burlington, VT	0.765
Charleston, SC	0.894
Charleston, WV	0.918
Charlotte, NC	0.854
Cheyenne, WY	0.826
Chicago, IL	1.037
Cincinnati, OH	0.904
Cleveland, OH	1.053
Columbia, SC	0.854
Columbus, OH	0.880
Dallas, TX	0.568

City	Cost Factor
Dayton, OH	0.895
Denver, CO	0.683
Des Moines, IA	0.922
Detroit, MI	1.105
Duluth, MN	1.021
El Paso, TX	0.707
Erie, PA	0.866
Fairbanks, AK	1.396
Fargo, ND	0.819
Grand Rapids, MI	0.754
Hartford, CT	1.236
Honolulu, HI	1.181
Houston, TX	0.822
Indianapolis, IN	0.827
Jackson, MS	0.517
Jacksonville, FL	0.841
Kansas City, MO	0.901
Knoxville, TN	0.637
Lansing, MI	0.862
Las Vegas, NV	1.042
Little Rock, AR	0.911
Los Angeles, CA	1.113
Louisville, KY	0.839
Madison, WI	0.831
Manchester, NH	0.765
Memphis, TN	0.729
Miami, FL	0.616
Milwaukee, WI	0.944
Minneapolis, MN	0.893
Montgomery, AL	0.596
Nashville, TN	0.663
Newark, NJ	1.287
New Haven, CT	1.294
New Orleans, LA	0.863
New York, NY	1.491
Norfolk, VA	0.560
Oakland, CA	1.478
Oklahoma City, OK	0.608

CITY COST INDEXES

City	Cost Factor
Omaha, NE	0.752
Philadelphia, PA	1.169
Phoenix, AZ	0.858
Pittsburgh, PA	0.899
Portland, ME	0.840
Portland, OR	0.887
Providence, RI	1.092
Raleigh, NC	0.862
Reading, PA	0.818
Richmond, VA	0.793
Rochester, NY	0.983
Sacramento, CA	1.042
St. Louis, MO	0.993
St. Paul, MN	1.049
St. Petersburg, FL	0.817
Salt Lake City, UT	0.887
San Diego, CA	1.175
San Francisco, CA	1.232
Santa Fe, NM	0.925
Savannah, GA	0.773
Scranton, PA	0.857
Seattle, WA	0.943
Sioux Falls, SD	0.756
South Bend, IN	0.818
Spokane, WA	0.852
Stamford, CT	1.272
Syracuse, NY	0.947
Tampa, FL	0.847
Toledo, OH	0.950
Topeka, KS	0.594
Trenton, NJ	1.155
Tulsa, OK	0.594
Washington, DC	0.816
Wichita, KS	0.820
Wilmington, DE	0.977
York, PA	0.764
Youngstown, OH	0.904

Chapter 17

MENSURATION

In this chapter the contractor or estimator will find all the basic math needed to estimate quantities and costs accurately and efficiently. In a number of these tables, the quantities are stated in decimals. Decimals make stating the fractional part of inches, feet and yards easier and more efficient, and because the use of calculators is pretty much universal these days, contractors and estimators must be thoroughly familiar with the decimal system.

Estimating is nearly all "figures" of one kind or another, so it is essential that the contractor and estimator possess a fair working knowledge of arithmetic. Most of the estimator's computations involve measurements of surface and cubical contents and are stated in lineal feet (lin. ft.), square feet (sq. ft.), square yards (sq. yds.), squares (sqs.), which contain 100 sq. ft., cubic feet (cu. ft.) and cubic yards (cu. yds.). Quantities are expressed in specific ways for particular items, and one will encounter units of measure such as thousands of brick, board feet of lumber and the like. But the basic units listed above are the most commonly encountered by the construction estimator.

The following abbreviations and symbols are standard and are used throughout this book:

	Abbreviation	Symbols
Inches	in.	"
Lineal feet	lin. ft., l.f.	'
Feet and inches	ft. in	2'-5"
Square feet	sq. ft., s.f.	
Square yards	sq. yds., s.y.	
Squares	sqs.	
Cubic feet	cu. ft., c.f.	
Cubic yards	cu. yds., c.y.	
Board feet	b.f.	

Lineal Measure Equivalents

12 inches	12 in.	12"	1 foot	1 lin. ft.	1'-0"
3 feet	3 ft.	3'-0"	1 yard	1 yd.	
16-1/2 feet	16-1/2 ft.	16'-6"	1 rod	1 rd.	
40 rods	40 rds.	1 furlong	1 fur.		

8 furlongs . 8 fur. .... 1 mile .... 1 mi.
5,280 feet . 1,760 yds. .... 1 mile .... 1 mi.

Square Measure or Measure of Surfaces

144 square inches .... 144 sq. in. .... 1 square foot ... 1 sq. ft.
  9 square feet .... 9 sq. ft. .... 1 square yard . 1 sq. yd.
100 square feet .... 100 sq. ft. .... 1 square* .... 1 sq.
30-1/4 square yards .... 30-1/4 sq. yds. . 30.25 sq. yds. .. 1 square rod   1 sq. rd.
160 square rods .... 160 sq. rds. ...... 1 acre .... 1 A.
43,560 square feet........... 43,560 sq. ft. .... 4,840 sq. yds. .. 1 acre .... 1 A.
       640 acres .... 640 A. .... 1 square mile .. 1 sq. mi.

*A measure commonly used by architects and builders.

Cubic Measure or Cubical Contents

1,728 cubic inches .... 1,728 cu. in. ...... 1 cubic foot ...... 1 cu. ft.
  27 cubic feet .... 27 cu. ft. .... 1 cubic yard .... 1 cu. yd.
128 cubic feet.................. 128 cu. ft. .... 1 cord .... 1 cd.
24-3/4 cubic feet .... 24-3/4 cu. ft. .... 24.75 cu. ft. ..... 1 perch* .... 1 P.

*A perch of stone is nominally 16-1/2 feet long, one foot high and 1-1/2 feet thick, containing 24-3/4 cu. ft. However, in some states, especially those west of the Mississippi River, rubble work is figured by the perch containing 16-1/2 cu. ft. Before submitting prices on masonry by the perch, find out the common practice in your locality.

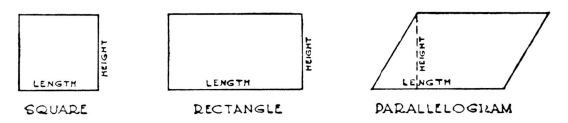

SQUARE   RECTANGLE   PARALLELOGRAM

MENSURATION

Area of a Square, Rectangle or Parallelogram. Multiply the length by the breadth or height. Example: Obtain the area of a wall 22'0" long by 9'-0" high. 22 x 9 = 198 sq. ft.

Area of a Triangle. Multiply the base by 1/2 the altitude or perpendicular height. Example: Find the area of the end gable of a house 24'-0" wide by 12'-0" high from the base to high point of roof. 24 x 6 = 144 sq. ft.

Circumference of a Circle. Multiply the diameter by 3.1416 (3-1/7). Example: Find the circumference or distance around a circle, the diameter of which is 12'-0". 12 x 3.1416 = 37.6992 feet. The circle is about 38 lin. ft. around.

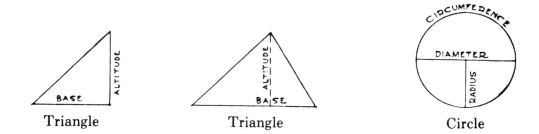

Triangle   Triangle   Circle

Area of a Circle. Multiply the square of the diameter by 0.7854, or multiply the radius by 3.1416. Example: Find the area of a round concrete column 24" (2'-0") in diameter. The square of the diameter is 2 x 2 = 4 lin. ft. 4 x 0.7854 = 3.1416 sq. ft., the area of the circle. The radius is 1/2 the diameter. If the diameter is 2'-0", then the radius would be 1'-0". The square of the radius is 1 x 1 = 1. 1 x 3.1416 = 3.1416 sq. ft.

Cubical Content of a Circular Column. Multiply the area of the circle by the height. Example: Find the cubical content of a round concrete column 2'-0" in diameter and 14'-0" long. From the previous example the area of a circle 2'-0" in diameter is 3.1416 sq. ft. 3.1416 x 14 = 43.9824 cu. ft. of concrete in each column.

Cubical Content of Any Regular Shape. Multiply length by breadth (or height) by thickness. Computations of this kind are used extensively in estimating all classes of building work, such as excavating, concrete foundations, reinforced concrete, brick masonry, cut stone and granite. Example: Find the cubical content of a wall 42'-0" long, 5'-6" high and 1'-4" thick. 42 x 5.5 x 1.334 = 308 cu. ft. To convert cubic feet to cubic yards, divide 308 by 27 = 11.41 cu. yds.

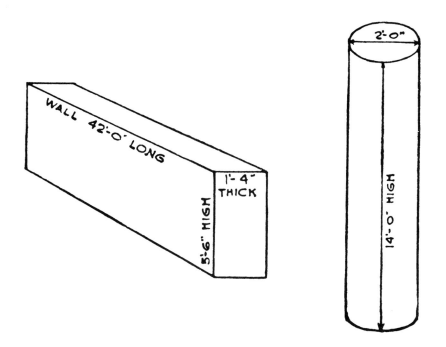

Using the Decimal System in Construction Estimating

In construction work nearly all figures are either feet and inches or dollars and cents. The decimal system is convenient in making many computations. All fractions have two components, a *numerator* and a *denominator*. For example, in the fraction 1/2, "1" is the numerator and "2" is the denominator. A decimal fraction, usually called simply a *decimal,* is a fraction whose denominator is some power of 10—10, 100, 1000, etc.

The denominator of a decimal fraction is not shown, because it is not necessary. We know that the decimal will be one tenth, one hundredths or one thousandths by the number of digits that appear to the right of the *decimal point.* For example, U.S. monetary amounts are always a good example of decimals. One half of a dollar is usually written as $0.50. We could right this decimal as 0.5, meaning "five tenths" but what is relevant is that it represents not five tenths of a dollar but rather fifty hundredths, or 50 cents. A good way to visual it as follows:

MENSURATION

0.5	means 5/10, for the	5 extends to the	10th place;
0.25	means 25/100, for the	25 extends to the	100th place;
0.125	means 125/1000, for the	125 extends to the	1000th place.

The names of the *decimal places* are:

Thou-sands	Hun-dreds	Tens	Units	(Decimal Point)	Tenths	Hun-dredths	Thou-sandths	Ten-Thou-sandths
1	3	4	5	.	2	7	6	5

This number is read "one thousand three hundred forty-five and two thousand seven hundred and sixty-five ten thousandths." The orders beyond the ten thousandths are hundred thousandths, millionths, ten millionths, etc. The number above is known as a *mixed decimal,* which is a number that includes both a whole number and a decimal.

When there is not a whole number in the decimal number, such as "0.5", it is not absolutely necessary to use the zero at the left of the decimal point, because 0.5 means the same as .5, but a zero is very helpful in calling attention to the decimal point. Without the zero, the decimal point can be easily missed.

Table of Feet and Inches Reduced to Decimals

The following table illustrates how feet and inches may be expressed in four different ways, all meaning the same thing.

1	inch =	1"	= 1/12th	foot =	0.083
1-1/2	inches =	1-1/2"	= 1/8th	foot =	0.125
2	inches =	2"	= 1/6th	foot =	0.1667
2-1/2	inches =	2-1/2"	= 5/24ths	foot =	0.2087
3	inches =	3"	= 1/4th	foot =	0.25
3-1/2	inches =	3-1/2"	= 7/24ths	foot =	0.2917
4	inches =	4"	= 1/3rd	foot =	0.333
4-1/2	inches =	4-1/2"	= 3/8ths	foot =	0.375
5	inches =	5"	= 5/12ths	foot =	0.417
5-1/2	inches =	5-1/2"	= 11/24ths	foot =	0.458
6	inches =	6"	= 1/2	foot =	0.5
6-1/2	inches =	6-1/2"	= 13/24ths	foot =	0.5417

```
7        inches =    7"      = 7/12ths foot  =   0.583
7-1/2    inches =    7-1/2"  = 5/8ths  foot  =   0.625
8        inches =    8"      = 2/3rds  foot  =   0.667
8-1/2    inches =    8-1/2"  = 17/24ths foot =   0.708
9        inches =    9"      = 3/4ths  foot  =   0.75
9-1/2    inches =    9-1/2"  = 19/24ths foot =   0.792
10       inches =    10"     = 5/6ths  foot  =   0.833
10-1/2   inches =    10-1/2" = 7/8ths  foot  =   0.875
11       inches =    11"     = 11/12ths foot =   0.917
11-1/2   inches =    11-1/2" = 23/24ths foot =   0.958
12       inches =    12"     = 1       foot  =   1.0
```

Example: Write 5'-7 1/2" as a decimal. The easiest way to do it, of course, is with an electronic calculator, which always works in decimals. To find the decimal equivalent of 7-1/2", divide 7-1/2 by 12, which equals 0.625. The decimal equivalent is 5.625 feet. We will forego a long discussion of why this calculation works. Suffice it to say that you divide by the number 12, because you are trying to establish the decimal equivalent in feet for the number of inches.

Table of Common Fractions in Decimals

These are decimal equivalents of common fractions frequently used in estimating.

1/16 .... 0.0625	7/16 .... 0.4375
1/8 .... 0.125	1/2 .... 0.5
3/16 .... 0.1875	9/16 .... 0.5625
1/4 .... 0.25	5/8 .... 0.625
5/16 .... 0.3125	11/16 .... 0.6875
3/8 .... 0.375	3/4 .... 0.75
13/16 .... 0.8125	15/16 .... 0.9375
7/8 .... 0.875	8/8 .... 1.0

MENSURATION

The following table illustrates how minutes can be reduced to fractional or decimal parts of an hour.

Number of Minutes		Fractional Part of an Hr.		Decimal Part of an Hr.	Number of Minutes		Fractional Part of an Hr.		Decimal Part of an Hr.
1	=	1/60	=	0.0167	31	=	31/60	=	0.5167
2	=	1/30	=	0.0333	32	=	8/15	=	0.5333
3	=	1/20	=	0.05	33	=	11/20	=	0.55
4	=	1/15	=	0.0667	34	=	17/30	=	0.5667
5	=	1/12	=	0.0833	35	=	7/12	=	0.5833
6	=	1/10	=	0.10	36	=	3/5	=	0.60
7	=	7/60	=	0.1167	37	=	37/60	=	0.6167
8	=	2/15	=	0.1333	38	=	19/30	=	0.6333
9	=	3/20	=	0.15	39	=	13/20	=	0.65
10	=	1/6	=	0.1667	40	=	2/3	=	0.6667
11	=	11/60	=	0.1833	41	=	41/60	=	0.6833
12	=	1/5	=	0.20	42	=	7/10	=	0.70
13	=	13/60	=	0.2167	43	=	43/60	=	0.7167
14	=	7/30	=	0.2333	44	=	11/15	=	0.7333
15	=	1/4	=	0.25	45	=	3/4	=	0.75
16	=	4/15	=	0.2667	46	=	23/30	=	0.7667
17	=	17/60	=	0.2833	47	=	47/60	=	0.7833
18	=	3/10	=	0.30	48	=	4/5	=	0.80
19	=	19/60	=	0.3167	49	=	49/60	=	0.8167
20	=	1/3	=	0.333	50	=	5/6	=	0.8333
21	=	7/20	=	0.35	51	=	51/60	=	0.85
22	=	11/30	=	0.3667	52	=	13/15	=	0.8667
23	=	23/60	=	0.3833	53	=	53/60	=	0.8833
24	=	2/5	=	0.40	54	=	9/10	=	0.90
25	=	5/12	=	0.4167	55	=	11/12	=	0.9167
26	=	13/30	=	0.4333	56	=	14/15	=	0.9333
27	=	9/20	=	0.45	57	=	19/20	=	0.95
28	=	7/15	=	0.4667	58	=	29/30	=	0.9667
29	=	29/60	=	0.4833	59	=	59/60	=	0.9833
30	=	1/2	=	0.5	60	=	1	=	1.0

Now find the labor cost for 4.6167 hrs. at $7.50 per hour.

```
        Hours       4.6167
  x Hourly Rate     $7.50
                   2308350
                   323169
    Total Cost    $34.625250 or $34.63
```

Similar Decimals. Decimals that have the same number of decimal places are called similar decimals. Thus, 0.75 and 0.25 are similar decimals and so are 0.150 and 0.275; but 0.15 and 0.275 are dissimilar decimals.

To reduce dissimilar decimals to similar decimals, give them the same number of decimal places by annexing or cutting off zeros. Example, 0.125, 0.25, 0.375 and 0.5 may all be reduced to thousandths as follows: 0.125, 0.250, 0.375 and 0.500.

To Reduce a Decimal to a Common Fraction. Omit the decimal point, write the denominator of the decimal and then reduce the common fraction to its lowest terms.

Example: 0.375 equals 375/1000, which reduced to its lowest terms equals 3/8.

$$25 \left| \frac{375}{1000} = 5 \right| \frac{15}{40} = 3/8$$

How to Add Decimals. To add numbers containing decimals, write like orders under one another, and then add as with whole numbers. Example, add 0.125, 0.25, 0.375 and 1.0.

The total of the addition is 1.750, which equals 1 750/1000 and which may further be reduced to 1 3/4

```
  0.125
  0.25
  0.375
  1.0
  1.750
```

How to Subtract Decimals. To subtract one number from another, when either or both contain decimals, reduce to similar decimals, write like orders under one another and subtract as with whole numbers.

Subtract 20 hours and 37 minutes from 27 hours and 13 minutes, both being written in decimals. The difference is 6.6000 or 6.6 hours, which reduced to a common fraction is 6 3/5 hours or 6 hours and 36 minutes.

```
27.2167
20.6167
 6.6000
```

How to Multiply Decimals. The same method is used as in multiplying other numbers, and the result should contain as many numbers to the right of the decimal point as there are in both of the numbers multiplied. Example: Multiply 3'-6" x 4'-9".

```
4'-9" = 4-3/4 ft. = 4.75 Multiplicand       4.75
3'-6" = 3-1/2 ft = 3.5 Multiplier            3.5
To multiply, proceed as illustrated         2375
                                            1425
Product                                    16625
                                         or 16.625
```

In the above example there are two decimal places in the multiplicand and one decimal place in the multiplier. The result or product should contain the same number of decimals as the multiplicand and multiplier combined, which is three. Starting at the right and counting to the left three places, place the decimal point between the two sixes. The result would be 16.625, which equals 16 625/1000 = 16 5/8 sq. ft.

Practical Examples Using Decimals

The quantities of cement, sand and gravel required for one cu. yd. of concrete are ordinarily stated in decimals. For instance, the proportions for a cu. yd. of concrete that is 1part cement, 2 parts sand, and 4 parts gravel is usually stated as 1:2:4. It requires the following materials:

1.50 bbls. cement = 1 50/100 bbls. = 1 1/2 bbls. = 6 sacks
0.42 cu. yds. sand = 42/100 cu. yds. = 21/50 cu. yds. = 11 1/3 cu. ft.
0.84 cu. yds. gravel = 84/100 cu. yds. = 22 2/3 cu. ft.

There are 4 sacks of cement to the bbl., and each sack weighs 94 lbs. and contains approximately 1 cu. ft. of cement.

There are 27 cu. ft. in a cu. yd.; to obtain the number of cu. ft. of sand required for a yard of concrete, 0.42 cu. yds. = 42/100 cu. yds. = 21/50 cu. yds. and 21/50 of 27

$$\text{cu. ft.} = \frac{21 \times 27}{50} = \frac{567}{50}$$

567 ÷ 50 = 11.34 or 11 1/3 cu. ft. sand

To find the cost of a cu. yd. of concrete based on the above quantities, and assuming it requires 2 1/4 hours labor time to mix and place one cu. yd. of concrete, proceed as follows:

1 1/2 bbls. cement =	1.50 bbls. cement @	$9.00 per bbl.	=	$13.5000
11 1/3 cu. ft. sand =	0.42 cu. yd. sand @	5.75 per cu. yd.	=	2.4150
22 2/3 cu. ft. gravel =	0.84 cu. yd. gravel @	5.75 per cu. yd.	=	4.8300
2 1/4 hours labor =	2.25 hrs. labor @	9.80 per hr.	=	22.0500
Cost per cu. yd.				$42.7950

Notice that the total is carried to four decimal places. This is to illustrate the actual figures obtained by multiplying. The total would be $42.80 per cu. yd. of concrete.

In making the above multiplications, you will note there are 2 decimal places in both the multiplicand and the multiplier of all the amounts multiplied, so the result should contain as many decimals as the sum of the multiplicand and multiplier, which is 4. You will note all the totals contain 4 decimal places. Always bear this in mind when making your multiplications, because if the decimal place is wrong, it makes a difference of 90 percent of your total. You know the correct result is $42.7950 but suppose by mistake you counted off 5 decimal places. The result would be $4.27950 or $4.28 per cu. yd., or just about 1/10 enough; or if you made a mistake the other way and counted off just 3 decimal places, the result would be $427.950 or $427.95 per cu. yd., or just about 10 times too much. Watch your decimal places.

The same method is used in estimating lumber. All kinds of framing lumber are ordinarily sold by the measure of 1,000 board feet, so 1,000 is the unit or decimal used when estimating lumber.

Suppose that you buy 150 pieces of 2"x8"-16'-0", which contains 3,200 b.f. This is equivalent to 3-200/1000 = 3-2/10 = 3-1/5 thousandths, or stated in decimals it may be either 3.2 or 3.200. The cost of lumber at $230.00 per 1,000 b.f. would be

MENSURATION

obtained by multiplying 3.200 ft. at $230.00 per 1,000 b.f. as follows: 3.200M b.f. of lumber @ $736.00000 or $736.00.

Note there were 3 decimal places in the multiplicand and 2 decimal places in the multiplier, so the result should contain 5 decimal places, but inasmuch as all the decimals are zeros, drop all but two of them to designate 736 dollars and no cents.

Things to Remember When Using Decimals

0.0	at the left of a figure indicates a decimal
0.6	indicates tenths, thus .6 = 6/10 = .60 = 60/100 = .600 = 600/1000
0.06	indicates hundredths, thus .06 = 6/100 = .060 = 60/1000
0.006	indicates thousandths, thus .006 = 6/1000
0.0006	indicates ten thousandths, thus .0006 = 6/10000

When multiplying decimals always remember that the result or product must contain as many decimal places as the sum of the decimal places in both the multiplicand and the multiplier. Find the cost of 3 hrs. 20 min. laborer time at $9.80 per hour.

```
3 hours 20 minutes = 3 1/2 hours= 3.333      multiplicand
9 dollars 80 cents per hour =       = 9.80   multiplier
                               2666640
                                299997
                               32666340      = 32.6666340 = $32.67
```

Be sure the decimal point is in the RIGHT place and the rest is easy.

Multiply	by	to obtain
acres	0.404687	hectares
”	4.04687×10^{-3}	square kilometers
ares	1076.39	square feet
board feet	144 sq in. $\times$ 1 in.	cubic inches
” ”	0.0833	cubic feet
bushels	0.3521	hectoliters
centimeters	3.28083×10^{-2}	feet
”	0.3937	inches
cubic centimeters	3.53145×10^{-5}	cubic feet
” ”	6.102×10^{-2}	cubic inches
cubic feet	2.8317×10^{4}	cubic centimeters
” ”	2.8317×10^{-2}	cubic meters
” ”	6.22905	gallons, Imperial
” ”	0.2832	hectoliters
” ”	28.3170	liters
” ”	2.38095×10^{-2}	tons, British shipping
” ”	0.025	tons, U.S. shipping
cubic inches	16.38716	cubic centimeters
cubic meters	35.3145	cubic feet
” ”	1.30794	cubic yards
” ”	264.2	gallons, U. S.
cubic yards	0.764559	cubic meters
” ”	7.6336	hectoliters
degrees, angular	0.0174533	radians
degrees, F (less 32 F)	0.5556	degrees, C
” C	1.8	degrees, F (less 32 F)
foot pounds	0.13826	kilogram meters
feet	30.4801	centimeters
”	0.304801	meters
”	304.801	millimeters
”	1.64468×10^{-4}	miles, nautical
gallons, Imperial	0.160538	cubic feet
” ”	1.20091	gallons, U. S.
” ”	4.54596	liters
gallons, U.S	0.832702	gallons, Imperial
” ”	0.13368	cubic feet
” ”	231.	cubic inches
” ”	0.0378	hectoliters
” ”	3.78543	liters
grams, metric	2.20462×10^{-3}	pounds, avoirdupois
hectares	2.47104	acres
”	1.076387×10^{5}	square feet

MENSURATION

Conversion Factors (Cont'd)

Multiply	by	to obtain
hectares	3.86101×10^{-3}	square miles
hectoliters	3.531	cubic feet
"	2.84	bushels
"	0.131	cubic yards
hectoliters	26.42	gallons
horsepower, metric	0.98632	horsepower, U. S.
horsepower, U.S.	1.01387	horsepower, metric
inches	2.54001	centimeters
"	2.54001×10^{-2}	meters
"	25.4001	millimeters
kilograms	2.20462	pounds
"	9.84206×10^{-4}	long tons
"	1.10231×10^{-3}	short tons
kilogram meters	7.233	foot pounds
kilograms per m	0.671972	pounds per ft
kilograms per sq cm	14.2234	pounds per sq in.
kilograms per sq m	0.204817	pounds per sq ft
" " " "	9.14362×10^{-5}	long tons per sq ft
kilograms per sq mm	1422.34	pounds per sq in.
" " " "	0.634973	long tons per sq. in.
kilograms per cu m	6.24283×10^{-2}	pounds per cu ft
kilometers	0.62137	miles, statute
"	0.53959	miles, nautical
"	3280.7	feet
liters	0.219975	gallons, Imperial
"	0.26417	gallons, U.S.
"	3.53145×10^{-2}	cubic feet
"	61.022	cubic inches
meters	3.28083	feet
"	39.37	inches
"	1.09361	yards
miles, statute	1.60935	kilometers
" "	0.8684	miles, nautical
miles, nautical	6080.204	feet
" "	1.85325	kilometers
" "	1.1516	miles, statute
millimeters	3.28083×10^{-3}	feet
"	3.937×10^{-2}	inches
pounds, avoirdupois	453.592	grams, metric
" "	0.453592	kilograms
" "	4.464×10^{-4}	tons, long
" "	4.53592×10^{-4}	tons, metric
pounds per ft	1.48816	kilograms per m

Conversion Factors (Cont'd)

Multiply	by	to obtain
pounds per sq ft	4.88241	kilograms per sq m
pounds per sq in	7.031×10^{-2}	kilograms per sq cm
" " " "	7.031×10^{-4}	kilograms per sq mm
pounds per cu ft	16.0184	kilograms per cu m
radians	57.29578	degrees, angular
square centimeters	0.1550	square inches
square feet	9.29034×10^{-4}	ares
square feet	9.29034×10^{-6}	hectares
" "	0.0929034	square meters
square inches	6.45163	square centimeters
" "	645.163	square millimeters
square kilometers	247.104	acres
" "	0.3861	square miles
square meters	10.7639	square feet
" "	1.19599	square yards
square miles	259.0	hectares
" "	2.590	square kilometers
square millimeters	1.550×10^{-3}	square inches
square yards	0.83613	square meters
tons, long	1016.05	kilograms
" "	2240.	pounds
" "	1.01605	tons, metric
" "	1.120	tons, short
tons, long, per sq ft	1.09366×10^{-4}	kilograms per sq m
tons, long, per sq in	1.57494	kilograms per sq mm
tons, metric	2204.62	pounds
" "	0.98421	tons, long
" "	1.10231	tons, short
tons, short	907.185	kilograms
" "	0.892857	tons, long
" "	0.907185	tons, metric
tons, British shipping	42.00	cubic feet
" " "	0.952381	tons, U. S. shipping
tons, U. S. shipping	40.00	cubic feet
" " "	1.050	tons, British shipping
yards	0.914402	meters

INDEX

A

Air cleaners, electronic 203
Air conditioning 216
   cost per installation 217
   labor to install 217
Alarms, household 203
Aluminum siding 269
  (see Siding)
Ampacity of insulated copper conduction 190
Asphalt shingles
   estimating quantities 231
   installation 232
   nails required 231
Attics, conversion 35

B

Base cabinets, removal 11
   replace 14
Baths, material costs 34
Bath fixtures 21
Bathtubs 182
Bathtubs
   removal 21
   roughing in plumbing 21
Bidet, install 20, 183
Brick, repairs 271, 272, 273

C

Cabinets
   installing bath 21
   installing kitchen 13
   removing kitchen 10
Carpentry, exterior finish 275

Cedar shakes
    estimating coverage 234
    grades and sizes 262
Ceilings
    to remove 121
    to install 121
Ceiling fans 203
Chimneys, to rebuild 274
Chimney flashing 229
Concrete, slabs 52
Concrete block, repairs 272, 273
Conduit electrical 186
Corner boards, to place 275
Cornices, to place 275
Cupolas, to place 276

D

Decks 312
    labor to install 313
Dishwashers 183
Doors 287
    bifold, to install 303, 305
    bypass sliding, to install 305
    exterior 295
    folding, to install 305
    hardware, labor to set 306
    installing 298, 301, 303
    interior 303
    labor to install 301
    sliding glass, to install 300
    sliding glass, to remove 300
Door openers, electric 301
Dormers, add 39
Drip edges, to install 229
Drywall
    required thicknesses 110
    taping 116
    to install 113

INDEX

Ductwork, install 13

E

Eaves flashing, to install 229
Electrical 185
    basements, attics 204
    distribution 196
    estimating, basic circuit 191
    grounding circuits 205
    installing service entrance 194
    labor per circuit, one connection 191
    labor per circuit, through studs 192
    material prices 206
    rewiring, labor to install 197
    to determine power needs 188
Excavating, sewer pipe 174

F

Flashing, roofing 229
Flooring
    estimating labor 91, 92
    estimating material 90
    hardwood, labor to install 92
    oak block 98
    parquet 97
    sanding 93-94
    staining/sealing 96
    tile and sheet 98
Flooring
    blocking 36
    removal 12, 89
    sleepers 44
Furring strips, masonry walls 44, 112

G

Garbage disposal, install 17, 183
Grounding circuits 205
Gutter and downspouts 241
    installation 245
    material costs 242
    removal 244

H

Hardboard siding 263
Hardware, labor to set 306
Heaters
    auxiliary, bath 24, 203
    electric 181
    gas 181
Heating systems 209
    costs, installed 215
    estimating costs 209
    estimating labor costs 219
    hot water 218
    labor to balance 215
    loads 210
    warm air 211

I

Insulation
    basements 45
    room additions 49

K

Kitchens, design of 5

INDEX

L

Lavatories
    removal 23
    installation 23, 182
Lighting
    low voltage 203
    outdoor fixtures 203
Lumber, estimating quantities 54

M

Masonry repairs 271
Moldings
    removal 123
    replace 123-129

N

Nails
    quantities and kinds 65
    for asphalt shingles 231, 234
    for hardboard siding 263

P

Paint
    covering capacity, interior 143
    estimating interior dimensions 140
    estimating labor 145
    hardwood floors 154
    walls and ceilings 155
Painting
    covering capacity, exterior 279
    estimating exterior dimensions 276
    estimating labor costs 280
    labor and material tables 284-286
    labor to burn off 280
    sanding and puttying 280

Parquet, to install 97
Pipe fittings, estimating quantities of 174
Pipe,
    gas, to install 213
    sizes and prices 177
    smoke, to install 213
Plaster walls 108
    combined labor and material costs 109
Plenum chambers, to install 213
Plumbing, estimating 173
    fixtures, labor placing 175
    labor roughing in 175
    preparing detailed estimates 175
Plywood paneling
    estimating materials for 111
    installing 113
Porch
    days to replace 308
    debris removal 307
    labor cost to repair 308
    labor to repair 308
    labor to wreck 307
    typical replacement costs 308-310

R

Range
    hoods, install 16
    tops, removal 11
Registers, heating 214
Roofing 223
    estimating material required 233
    install underlayment 228
    labor to install asphalt shingles 234
    roll, to install 239
    to measure roof size 224-228
Rooms
    additions 49
    conversions 35

S

Sanding and puttying 281
Sheathing, labor to apply 53
Shingles and shakes, wooden
    roofing 236
        application for different slopes 238
        estimating quantities 237
        to install 238
    siding 259
        grades and sizes 261
        labor to install 260
        weather exposure 260
Shower stalls 182
Shut-off valves, install 11
Siding
    aluminum 269
        installing 269
    hardboard 263
        installing 263
    vinyl
        installing, horizontal 268
        installing, vertical 269
    wood 251
        beveled siding, labor placing 258
        beveled siding, quantities required 256
        drop siding, labor placing 259
        drop siding, quantities required 256
        ship lap, quantities required 257
        T&G, quantities required 257
Sink
    installing 16, 183
    removing 10
Skylights, installing 37
Soffits 274
    installing 16, 274

Stairways 77
   custom, spiral and disappearing 84
   labor to construct 78-83
   removal 78
   types of 77
Sub-flooring
   removing 89
   replacing 89

T

Taping, drywall 116
Tile, installing 25-34, 117
Toilets
   installing 20, 182
   removing 20

V

Valley flashing, installing 230
Vanities
   installing 23
   removing 23
Vent pipe flashing, installing 230
Ventilators, bath 23
Vinyl siding 266

W

Wall cabinets, hanging 13
Wall cover
   canvas 170
   flexwood 170
   vinyl 171
Wallpaper, hanging 167
Walls
   labor to frame 52
   new partition 37
   replaster 108

INDEX

stripping 107
Warm air ducts 213
Waterproofing 42
Weather exposure, shingles and shakes 260
Window
   styles 287
   types 290
Windows
   close opening 292
   installing 36, 292
   removing 291
Wiring, removal 187